SOLDERING AND BRAZING

SOLDERING AND BRAZING

Tubal Cain

Special Interest Model Books

Special Interest Model Books Ltd
P.O. BOX 327
Poole
Dorset
BH15 2RG
England
www.specialinterestmodelbooks.co.uk

First published by Argus Books Ltd. 1985
Reprinted 1987, 1988, 1990, 1991, 1994, 1995, 1997, 1999

This edition published by Special Interest Model Books Ltd. 2002

Reprinted 2003, 2005, 2007, 2008, 2010

ISBN 978-085242-845-0

Printed and bound in Malta by Melita Press

Contents

Foreword

Both soldering and brazing are very old crafts — many thousands of years old — and unlike many such they have persisted, using very similar basic techniques over the centuries. This may be the reason why the model engineer — and, for that matter, the 'jobbing professional' — is so ill-served in the literature. Such books as do exist were either written so long ago as to be out-of-date, or they deal with the subject from the point of view of the production engineer, concerned with the joining of perhaps tens of thousands of identical parts per week. Equally unfortunate is the fact that many of the developments in recent years — and there have been many — are related to and described in terms of automatic brazing machines, the robot, or soldering machines costing more than a Prime Minister's salary. Industrial soldering today is the province of the printed circuit board, and the car radiator or tincan manufacturer, and much of the brazing done in industry uses sophisticated heating processes and computer control. In very few books and papers does the 'manual' operator receive any consideration, and even when otherwise never more than a paragraph or two.

This book is written specifically for the manual operator, whether professional or amateur. I have tried to sort out the basic essentials from the mass of research literature so that you can see WHY the capillary joint gap is so important, HOW the joint action of flux and alloy conspires to make the bond, and — perhaps more important than is generally realised — how the joint design can influence both. It always helps to know the reasons behind the manipulation of the tools you are using. I have had great help from the manufacturers of both soft solder and brazing alloys and as a result have been able to offer some guidance through the proliferation of types, grades and specifications. In this connection I have, wherever possible, used the current British Standard nomenclature and I hope that you, too, will try to do the same when talking about the material you use. Specify BS 1845 AG21, and you will get the same material all over the world, but 'Jointfabwerkegesellshaft Schlippenflow 15' is likely to be obtainable only in Ruritania, and often only with some difficulty even there. Further, I would urge you to avoid the use of the term 'silver solder', for this CAN be confusing. It can describe alloys with melting points ('solidus') from 180°C up to 1200°C, and embrace materials containing metals as diverse as lead and palladium. The recommended terminology is—'Soldering' or 'Soft Soldering' when the working temperature lies below about 400°C, perhaps with the addition of 'silver bearing' to describe the solder when appropriate, and 'Brazing' or 'Hard Soldering' when the working temperature is higher than that – again with the appropriate prefix when describing the alloy. Being old-fashioned I find that I myself use the term 'Braze' when I am using spelter or plain brass as the filler alloy and 'Silver-braze' when using silver-bearing alloys.

I have done my best to cover all the general types of soldering and brazing you are likely to meet, and almost all the illustrations are from fabrications I have made or from 'exercise pieces' specially devised to illustrate the point. However, I have, throughout, concentrated on the *making of joints* (which is the important part) rather than on the finished article. Only where the nature of the fabrication may affect the making of the joint have I felt it necessary to go further. For this reason I have not dealt in detail with locomotive boiler-making. The loco boiler is large (and heavy) and it does require a fairly high capacity burner to deal with it, but compared with many smaller fabrications all the joints are simple and the procedures I have outlined will cover them. I have, in my time, made a number of boilers, locomotive and traction engine types included, but the really difficult brazing jobs have been quite small ones by contrast.

I have, throughout, had the 'beginner' or the relatively inexperienced reader in mind, and if the more expert reader finds at times that I am 'stating the obvious' I would simply ask him (or her) to remember the first time they handled a torch or soldering iron! However, I believe that there is a considerable body of information in the following pages which will be of interest, if not of value, to all. I can, perhaps, emphasise this point by saying that I know far more now than I did when I started to write, and am now making far better joints!

The chapter on 'Joint Design' gave me far more difficulty in the writing than any of the others. I cannot tell you HOW to design a joint, for they serve so many different functions, and any design is of necessity constrained by the nature of the component being made. In the end it seemed best to start by outlining the principles, then to cover the most common types of joints, and to complete the picture with a number of 'case histories', each chosen to illustrate one or more design factors. On the important question of workshop safety I have taken a different approach, by outlining in general the hazards and remedies in the chapter devoted to that subject, but providing a more specialised analysis as an appendix. This is a matter to be taken seriously, but we must retain a sense of proportion; most people die in bed, but this is a poor reason for sleeping on the floor!

Also in the appendix you will find data on the three most common fuel gases, and tables giving comprehensive details of brazing alloys under their commercial names. I am deeply indebted to the Calor Gas Company, and to Messrs Fry's Metals, Ltd., Johnson Matthey & Co. Ltd., and the Sheffield Smelting Co. Ltd (now Thessco Ltd.) whose staff have been generous in their help. I am also grateful to the British Standards Institution, for permission to reproduce data from their publications.*

I will conclude this foreword by making a point which is reiterated several times in the text. Soldering and brazing have established themselves as sound and reliable methods of joining metals over a period of several thousand years. Properly made joints can be expected to last for centuries – indeed, in many cases it is the parent metal which proves to be the more vulnerable to the ravages of time. Both 'soft' and 'hard' solders each have their proper place, each has its own virtues, and both have the right to be considered as 'sound engineering practice'. I hope that this book will help you to make joints which the archaeologists of the future will find worthy of their approbation.

Tubal Cain
September 1984

*Copies of all British Standards can be had from the British Standards Institution, Linford Wood, Milton Keynes MK14 6LE.

Chapter 1

Introduction

The accepted definition of both soldering and brazing is '. . . the joining of metals using a filler metal of lower melting point than that of the parent metals to be joined . . .' This is true enough, but leaves much unsaid. 'Bronze-welding' for example, uses just such a filler-rod, but is NOT 'Brazing'. To get to the bottom of the matter, let us compare the three processes of 'Welding', 'Glueing' and 'Soldering', noting in the bygoing that the only real difference between soldering and brazing is the melting-point of the filler material used. (The distinction was much clearer in the old days when the two processes were known as 'soft' and 'hard' soldering).

In Fig. 1 at 'A' we have a fusion welded joint. The two parts are united by means of the fillets 'w', which are made up of the same material as the parent plates. The latter have been melted locally, so the plates and fillets are literally one piece of the same material (though it should always be remembered that these fillets will be in the *cast* condition, whereas the plates themselves may be rolled or forged). The strength of the joint depends on the area of contact at the fillets, and in a butt joint as at 1B the parent plates have been 'prepared' by bevelling the edges, both to ensure complete penetration of the weld metal (absent in 1A) and to increase the effective contact area.

In Fig. 2 we have a glued joint. There is no fillet, the glue being disposed *between* the mating parts beforehand. Again, the strength will depend on the area in contact and, of course, on the strength of the glue and its bond to the parent material. With glues and cements this bond is usually due to the filler material – the glue – engaging with the surface irregularities of the joint faces; hence the need to roughen the surfaces. However, the important point to note

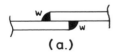

(a.)

(a) Welded lap joint.

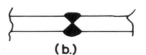

(b.)

(b) Welded butt joint, with

Fig. 1

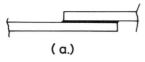

(a.)

Glued joints.

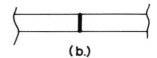

Fig. 2   *(a) Lap.   (b) Butt.*

(b.)

for our comparison is that the glue is set in place on the surfaces *before* they are brought together and that as a rule these parts must be clamped together until it sets. Some glues set by chemical action (the 'epoxy' type for example), some by evaporation of a solvent, and others are applied hot and set as they cool — resembling a 'solder' in that respect.

Fig. 3 shows a soldered joint, and at first sight it differs not at all from that of the glued joint in Fig. 2A. The difference, and it is a crucial one, is that unlike the glue the solder has been applied to the joint from the edge, and has penetrated the joint line by CAPILLARY ACTION. This is the crux of the matter. In 'welding' the filler material is deposited drop by drop onto the molten

Fig. 3.

'Capillary' joints, soft soldered or brazed.

surface of the joint. In a 'glued' joint the glue is applied before the two parts are united. In both 'soldering' and 'brazing' the filler penetrates the joint area from outside the mating surfaces. This is not to exclude altogether the setting of filler material between the parts before heating (a process known as 'sweating') but even when this *is* done the flow of the molten filler is the result of capillary forces.

These forces are considerable, and the filler (whether soft solder or brazing

alloy) can literally climb uphill. In Fig. 4 I have bent a piece of tinplate to a VEE shape, which was then heated up to the melting point of soft (tin-lead) solder. The solder was applied to the base of the crevice formed by the bend and then, when it had set, the bend has been torn open to reveal the solder filling. You will see that there is a

Fig. 4 *Showing that soft solder can 'climb' up a narrow gap.*

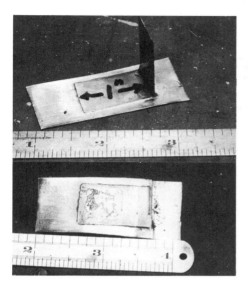

Fig. 5 *Flow through a capillary gap by soft solder. Top, (a) The lap joint, 'fed' from the right. (b) The same joint, torn open.*

marked increase in height as the gap between the two sides of the vee diminished – in this case the solder climbed up 1½ inches. In Fig. 5 I have made a lap joint. The soft solder was laid against the upright at the right-hand end and the joint gently heated from below until it melted. The joint is shown torn open at 5B, and you can see that the solder has flowed right through the joint. There is a small patch which was not properly wetted (we shall come to this point later) but this was due to the fact that the mating part was not thoroughly cleaned at that point. (The 'parent metal' in this case was cut from the can of one of Mr. Heinz' 57 varieties) Fig. 6 shows an attempt to make a similar demonstration using a brazing alloy melting at around 630°C. A groove of tapered depth ranging from 0.002 to

0.012 inch deep was cut in a piece of mild steel, and a small hole drilled at the blind end. A second piece of flat steel was clamped to it, with the groove filled with flux, and the whole heated up to brazing temperature. Brazing rod was then applied through the hole with the test-piece set vertically. The rod was fed in until it would take no more. After all had cooled the cover-plate was milled away, and the brazing alloy revealed. You will see that in this case the alloy climbed the full height of the test-piece – about 2.8 inches. It did **not** climb up at all, however, on the wide side of the groove on the right, another point of importance we shall return to later. There are a few 'inclusions' in the joint,

Fig. 6 *Climbing action of a silver-brazing alloy.*

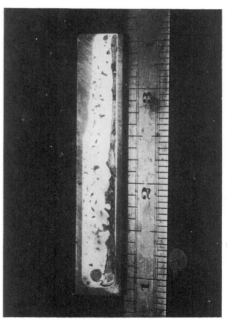

Fig. 7 *Wetting action of solder on brass. Left, no flux. Centre, using resin. Right, a semi-active flux.*

as you can see; these are my fault; the surface was left 'as milled' and I did not take enough care over cleaning before applying a water-based paste flux. However, the sample does show that the capillary attraction is very strong and can be relied upon to carry the alloy well into any joint provided that the joint gap is reasonable.

Wetting Capillary action can occur only when the fluid, whatever it is, wets the surface. This means that the nature of both the parent metal and the filler alloy will have an effect on the performance. Fortunately the manufacturers of solder and brazing alloys have tackled this problem for us, and it is rare to find a base metal which cannot be soldered or brazed, though some may be very difficult. However, there is a great deal of difference between their laboratories and your workshop, and the pristine surface which they may have used will seldom be found in practice. It is *imperative* that the surfaces be clean if proper wetting is to be achieved, and even more so if a proper bond is to be made. (Bonding is dealt with later in the chapter.) The most common obstacle is the oxide film which forms (surprisingly quickly) on almost all metal surfaces. To overcome this a 'flux' is used which, at the temperatures used in the process, will attack and remove any reasonable oxide film.

These are dealt with later in detail, and it will suffice to say now that the flux should *not* be used to clean up a dirty joint surface — it has enough on its hands in preventing the filler metal and joint surface from oxidising at the jointing temperature. The surfaces should be as clean as can be managed before starting work, and as we shall see later the cleaning method used can have quite an effect on the integrity of the joint.

You can check this point very easily for yourself — and, incidentally, compare the effectiveness of fluxes, too. Cut a strip of clean brass about 1 inch wide x 3 inch long and thoroughly clean the surface, preferably *not* with emery; use pumice powder and water, finish by washing with detergent and hot water, and then air drying it. Coat the centre third of the length with a paste flux, and one end with whatever other flux you have available. Take care to get no flux at all on the other third of the length. Cut off three small pieces of soft solder (*not* the resin cored stuff) and set one in the middle of each third of the length. Now heat the strip gently and evenly from underneath until the solder melts. You will see for yourself that even though the one end was really clean the solder has not spread at all — it does not 'wet' the surface. The effectiveness of the other fluxes used can be compared by the amount of 'spread' of the solder lump. I have done this in Fig. 7, with the unfluxed section on the left, pure resin used in the centre, and a proprietary paste flux on the right.

Bonding We have seen that 'glue' acts by getting a grip on the surface irregularities of the parent materials. This *can* happen with soldering and brazing too; the little lump of solder on the left of Fig. 7 was quite firmly stuck on, but it parted from the brass quite simply when given a sideways tap. Examination of a properly soldered joint when torn apart shows quite a different state of affairs. Indeed, most of you will have had the experience of trying to remove solder when it has 'got where it shouldn't have'! Somehow it almost seems to be necessary to go right below the surface of the original metal. The bonding effected in both hard and soft soldering is a *metallurgical* process. It is not necessary to go into this in any detail – indeed, it would be difficult to do so in a book this size; but a few words may help you to understand what is going on and, more important, to give you some idea of what *may* have happened when things don't go quite as they should.

The basis of all 'solders', hard or soft, is one or more 'active metals', which can form a type of alloy with the parent metals to be joined. Tin is one of the most active and is the basis for almost all soft solder. Copper, silver, zinc and many others show this form of activity, some quite general, others with greater affinity for one parent or base metal than others. The manufacturer of the solder or brazing alloy will have carried out extensive research to ensure that the composition of 'general purpose' fillers have a reasonably wide applica-

Fig. 7A *Erosion of a screw-in bit.*

tion, and at the same time will have developed special alloys to deal with the more difficult joints. There are few base metals (using the term to indicate 'the metal to be jointed') which cannot be brazed or soldered.

The active element in the filler forms an ALLOY with the base metal during the heating period. This may be surprising to those who have been told that alloys are formed by melting two (or more) metals together. However, again a little observation will show that it is possible for an alloy to form between a hot liquid and a solid. I suppose that everyone who has used a soldering iron will at some time or another have found it necessary to dress up the business end with a file, and will have found that once the solder itself has been removed there seems to be a hard coating on the bit. This is a tin-copper alloy, which has formed during previous use of the tool. It is quite marked. You may have found, too, that the end of a soldering iron bit seems to 'dis-

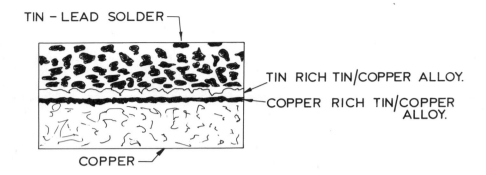

TIN – LEAD SOLDER

TIN RICH TIN/COPPER ALLOY.

COPPER RICH TIN/COPPER ALLOY.

COPPER

Fig. 8 *Alloy formation between tin/lead solder and copper. (After Thwaites).*

solve away' (see Fig. 7A). This again is due to the action of the tin in the solder on the copper, and some commercial soldering irons are 'iron plated' over the copper to reduce this. So, it can work both ways: copper being dissolved by the tin, and tin being alloyed with the solid copper of the bit.

The 'bond' between the base or parent metal and the filler is, therefore, an alloy layer as shown in Fig. 8. The bond is not mechanical, as in a glue, but metallurgical, and is very strong indeed. If the joint is close enough the joint may be almost entirely 'alloy', but it is seldom that a gap as close as this will permit the necessary capillary movement of the filler. It *can* happen, however, in a sweated joint. Incidentally, this alloy formation is one explanation of the fact that it seems to need a higher temperature to 'unbraze' a joint than it does to make it! Note that this alloy must not be confused with 'Intermetallic Compounds' sometimes referred to in articles on soft soldering. These do occur, but need an electron microscope to see them, and they are usually regarded as deleterious rather than forming part of the bond.

Conclusions From what I have said it will be seen that the basic principles of a soldered or brazed joint are (1) that the bond is the result of the formation of an alloy between one or more of the constituents of the filler material and the base or parent metal. (2) That the filler penetrates the joint by capillary action. These are the important points – the fact that the filler has a lower melting point than the base is only a matter of convenience; it would be very difficult to carry out the process if the reverse held true!

These principles lead to certain consequences, the understanding of which is imperative if good joints are to be made. (Or any joint at all, for that matter.) *First*, as the joint is a 'capillary' **THERE MUST BE A GAP.** It is only possible to make a brazed or soldered joint *without* an initial gap in certain specialised applications (e.g. 'sweating') which I shall deal with when the time comes. *Second*, the gap must lie between fairly well-defined limits. Too narrow, and the alloy will not flow; too wide, and the surface tension, which causes the capillary flow, will be insufficient. These limits are fairly wide for soft (tin-lead, or tin-lead-silver) solders, but smaller for the 'hard' or brazing solders. *Third*, capillary flow cannot occur unless the alloy can 'wet' the surface. The filler alloys are compounded by the makers so that this will occur, but the preparation of the base surfaces is entirely in the hands of the user; cleanliness is important. *Fourth*, the bond results from the formation of an alloy between filler and base, and this cannot form if any oxide layer intervenes. To prevent this we must use a flux which either destroys the oxide or prevents it from forming. (You may hear of the 'fluxless brazing' of copper from time to time. This is a misnomer; true, no flux is ADDED to the joint, but in fact the filler alloy contains a substance, usually phosphorus, which acts as a flux once the filler is molten.) Other important matters, such as the temperature at which the joint is made, the formation of fillets and so on, all spring from these four requirements, and will be dealt with later on.

However, I think I must deal with one other point now — nothing to do with what has gone before. Both soft soldering and brazing are jointing processes which have their own special merits. Welding is not 'better' than either, unless the job in question is such that welding is 'appropriate'. Despite the relatively high temperatures used in brazing (600-700°C) thermal distortion can be much less, and the process can be applied to workpieces so delicate that welding would be impossible. Similarly, soft soldering (at temperatures around 200°C) is not necessarily inferior to brazing, and has several advantages; the joint is easily undone if need be, thermal distortion is almost entirely absent, and the cost is very low. Which process is 'best' depends entirely on the application — and, perhaps, convenience.

Readers should beware of being influenced too much by 'the professionals'. Their advice is always to be welcomed, but it is only prudent to remember that what is appropriate in a commercial undertaking may not always be relevant to model making, and procedures which can pay their way in flow production plants may well prove to be hopelessly uneconomic in a jobbing workshop. Horses for courses, always!

Chapter 2

The Characteristics of Filler Metals

I use the term 'Filler Metal' to indicate the brazing alloy, spelter, or solder and in contrast to the 'base' or 'parent' metal which is being jointed. This is a general term and can be applied either to 'solder' (low melting-point material) or to brazing alloys of any sort. Where my remarks apply only to one or other of these I shall use the term 'solder', 'spelter' or 'braze metal' as is appropriate. I hope that this will avoid confusion.

It is possible to use pure metals as fillers, and for special reasons in the past I have often 'soldered' with pure tin, melting point 232°C. Pure copper (M.P. 1083°C) is used increasingly in industry for the fabrication of steel components. In fact, almost any of the 'active' metals can be used as jointing fillers. The almost universal use of alloys arises from several considerations. (1) *Cost*. Tin is very expensive, and except in a few situations the addition of a 'padder' material will effect an economy with no serious reduction either in strength or capillary action. (2) *Application temperature*. The addition of alloying elements reduces the melting temperature. We shall be dealing with this in detail shortly, but it is worth noting now that the addition of a more expensive alloying element can and often does reduce the overall cost of the joint, due to the reduced heating costs. (3) *Improved flow*. Quite small additions of alloying elements can effect a marked improvement in 'capillarity' — the cost may be somewhat increased but the joint is considerably improved. (4) *Mechanical properties*. Strength is but one of these. Resistance to corrosion, ability to resist higher temperatures, shock, or fatigue are also of importance. It must always be remembered that the filler metal is, within the joint, in the 'as cast' condition and the physical properties specified by the manufacturers apply to this condition.

This almost universal use of alloys as filler metal has very important consequences both for the designer of the joint and for the chap who is making it, and it is, unfortunately, true that many users do not fully understand the implications. At first sight it might appear reasonable to suppose that if we made an alloy from equal proportions of two metals the melting point would lie more or less midway between that of each component. This is far from being the case — indeed, for almost all alloys (certainly all used in metal joining) a proportion of the two alloy metals will be found which has a melting point below that of either. At other proportions of the two constituents you will find that there is no well defined melting point at all; the metal starts to melt, but remains 'pasty' until a higher temperature is reached when the whole becomes molten. In some cases this can be a nuisance, but in others ('wiped' joints for lead pipes, for example) is a positive advantage. The effect can best be seen from a melting-point diagram.

Eutectics and Melting Ranges. Fig. 9 shows the effective melting points of an alloy of tin and lead – soft solder, in fact. I am using this as an example because it is not untypical of most two-metal ('binary') alloys, and I have simplified the diagram to show only the essential points. The bottom

scale shows the percent of *tin* in the alloy, and that at the top of the diagram – running the other way – the percentage of lead. The left-hand scale is temperature in degrees C. It will be seen that pure lead (left hand end) melts at 327°C and pure tin at the right, at 232°C. Above the line ABC the alloy is fully molten whatever the composition. Below the line (almost straight) ADBEC it is always fully solid. Between the two lines the alloy will be more or less 'pasty' – a mixture of solid crystals swimming in molten material.

You will notice that the two lines coincide at B. This is the *only* proportion of a tin-lead alloy which has a definite melting temperature, at 183°C; it melts or solidifies almost instantaneously at this temperature. This particular composition is called the *EUTECTIC* alloy, and consists of 61.9% tin, 38.1% lead. At all other compositions

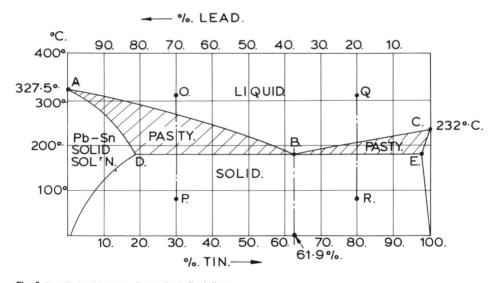

Fig. 9 *Simplified melting-range diagram for tin-/lead alloys.*

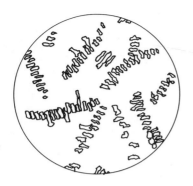

Fig. 10 *Sketch of a microscope view of 80% tin-20% lead alloy. The tin-rich crystals are surrounded by 'eutectic' material containing 61.9% tin.*

there is a *melting range* during which the metal is pasty. The line ABC is called the 'LIQUIDUS', being the temperature above which the alloy is wholly liquid, and ADBEC is the 'SOLIDUS', below which the metal is wholly solid. You will see that this solidus coincides with the eutectic temperature and this is characteristic of all simple alloys; the melting point of the 'eutectic' composition is also the 'solidus' for all compositions of the system.

Fig. 11 *A similar micrograph to Fig. 10 but for a 30% tin alloy. The dark patches are lead-rich alloy, again surrounded by the eutectic material.*

Now look at the line RQ, which represents the heating of an alloy containing 80% tin. In the solid state the metal looks something like Fig. 10 under the microscope. The christmas-tree shapes are crystals of tin-rich alloy and are surrounded by a matrix of eutectic material, 61.9% tin. There is an excess of tin in the alloy and this has separated out. As we reach the solidus temperature (DBE) the eutectic mass melts, but the tin-rich crystals stay solid. These begin to melt as the temperature rises still further until finally, when the liquidus temperature is achieved, they are all liquid. The reverse situation holds on cooling; the tin-rich parts solidify first, the mass gets less fluid as the temperature falls, until we reach the solidus line when the main body of eutectic alloy solidifies.

The line PO indicates a solder containing 30% tin. We now have an excess of *lead* over the eutectic composition. The solid metal would look like Fig. 11, with conglomerations of lead surrounded by the eutectic – not much of the latter, as there is so little tin present. Again, on heating this eutectic melts first and holds the lead-rich crystals in suspension as a pasty mass. Not until the liquidus line is reached will these melt.

So, in these two alloys we have a well-defined MELTING RANGE. From 183°C up to just over 200°C for the 80% tin alloy and from 183 up to 260°C for the 30% tin. This applies to all 'non-eutectic' alloys though you will appreciate that the shape of the melting curve (it is called an 'Equilibrium Diagram', for what that is worth) will be different for different materials. Many eutectics occur at very small proportions of one of the constituents, down at 5% or so. When you start looking at more elaborate alloys the diagram gets very complicated indeed. Some 'silver solders' or brazing alloys may have three or four constituents and one, Ag18 of BS 1984, has six. The eutectic temperature of these silver-bearing brazing alloys can be as low as 600°C, or as high as 960°C, and the melting range – the width of the pasty stage – may be quite short at 10°C for Ag2 or as much as 90°C in other cases. We shall deal with these differences later, but there are some general consequences of considerable importance which we had better look at first.

Liquation This is the name given to the situation where part of the alloy is fluid and the rest solid – the pasty stage. This *need* not be a nuisance. Indeed, in the tin-lead solders a special alloy BS 219 grade 'J' (with 30% tin, a little antimony and the rest lead) has a very long pasty stage and is much used for filling dents in car bodies. If maintained at between 190 and 210°C it can be worked almost like putty. However, when making a capillary joint the wide melting range CAN be a nuisance and where such an alloy is used it may be necessary to adopt special techniques. The problem is illustrated in Fig. 12. Here we have a lap-tee joint which is being heated from a flame or soldering bit and you will see

that just below the heat the joint temperature is 250°C, well above the liquidus of normal solder. However, at the other end of the joint, where the solder has been applied, the temperature is barely above the eutectic (which would be the solidus of the alloy). The result is that the eutectic part of the mass of solder will melt and flow into the joint, but will leave behind an unsightly solid lump of lead-rich material. The joint may well fill with solder, and a fillet will apparently be adding to the strength, but in fact the joint will not achieve its design strength – there would have been a good reason for the designer specifying a non-eutectic solder.

This situation can be even more serious in the case of silver-brazed joints, especially when 'step brazing' – that is, successive brazing operations on the same workpiece, a process which clearly needs a lower 'melting point' for each brazing operation. The alloy used for the first braze is likely to have a fairly wide melting range, and if 'liquation' occurs the joint will have been made with none but the lower melting point fraction of the alloy. When the next operation is started we do run the risk that this first joint may weaken or even fail. There are means of avoiding these problems – the applica-

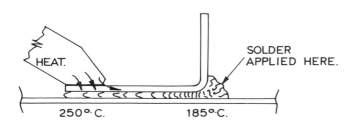

'Liquation'. The mouth of the joint is below the true liquidus temperature, so that the eutectic material only has penetrated the joint, leaving a lead rich mass at the entrance.

HEAT.

SOLDER APPLIED HERE.

250° C.   185° C.

Fig. 12

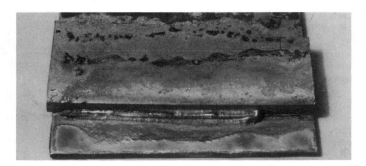

Fig. 13 *A silver-brazed joint, heated as in Fig. 12. The copper-rich part of the alloy has not melted.*

tion of sufficient heat to effect rapid melting of the alloy is the obvious one – but the effect must always be remembered, both when designing the joint and when actually brazing it up.

In Fig. 13 I have tried to made a 'bad' joint (not as easy as you would think when you want to do it on purpose!) to show the effect. A couple of pieces of scrap copper were properly cleaned and well fluxed, and a ½ inch length of brazing alloy set against the lap of the joint. This alloy has a liquidus of 780°C and a solidus at 740°C, containing 24% silver; one of the cheaper alloys used as a rule for step brazing. It was heated up to the working temperature relatively slowly, but not so slowly that the flux was overworked. You will see that there is a ridge of alloy standing a little away from the lap, with a small lump at one end. This is what is left of the piece of brazing rod; the joint is, in fact, filled, but the residue left behind contains much of the copper originally present in the alloy. There IS a 'brazed joint' but it will be of indeterminate strength and unknown composition. Fig. 14 shows a piece of the same alloy which has been fluxed and laid on a piece of scrap brass and then heated in the same way. You can see fairly clearly where some of the eutectic has flowed out of the strip of alloy, leaving a copper-rich component of much higher melting point behind. In this case I later refluxed and reheated until the residue also was melted. To do this the base metal had to be brought up to bright red (well over 850°C, far above the liquidus of the original alloy)

Fig. 14 *Liquation. Another example. This piece of brazing rod has been heated slowly, and only the eutectic part has melted.*

before the residue even offered to melt. For comparison, Fig. 15 shows a sample joint I have made using the identical alloy but this time the workpiece (two pieces of scrap) was brought well up to the operating temperature before applying the rod, which then melted quickly and as one and flowed into the joint. You will notice that there is a small fillet — characteristic of the brazing alloys which have a wide temperature range. This last example will, I hope, show that though liquation is a matter always to be borne in mind it is not a formidable obstacle. It does cause some practitioners apprehension when the workpiece is large or massive, but this is usually due to impatience! The work takes a long time to come up to temperature, and the temptation to apply the alloy too soon is irresistible. The remedy is obvious!

Conclusion Whilst some pure metals can be used as fillers these need not concern the average small user or the model engineer; either they are too costly for the advantage they give, or they need special techniques. All fillers we use will be alloys, and all will exhibit the eutectic phenomenon described in this chapter. All save the actual eutectic compositions will exhibit a 'range of melting', and a little experience enables

Fig. 15 *A specimen joint made with the same alloy as in Figs. 13 & 14, but heated to a temperature which ensured that the whole of the alloy was melted. A good joint is made.*

the practitioner to take advantage of this — it is a nuisance **only** in certain conditions, which I hope will now be recognisable. We shall be looking at these difficult conditions later on, but I will mention an important one now. When using alloys with long melting ranges, especially soft solder, it is necessary to allow sufficient time for the filler metal to set before disturbing the joint by handling it. Any disturbance during the pasty stage will result in a weak and porous joint. The special problem of 'liquation' can be overcome by getting enough heat in the right place at the right time.

Chapter 3

Soft Solders

Although 'soft' solder has been used for a very long time soldered joints were, until relatively recently, the province of the tinsmith, tin can maker, and, perhaps, the car radiator manufacturer. However, there has been an almost explosive volume of research done in the last ten years or so, both into the behaviour of the existing solders and in the development of new ones. This is largely due to the growth of the electronics industry. The 'Dolby' circuit in a small tape recorder needs 180 soldered connections and this is but part of the whole, and even quite humble domestic appliances now include some form of printed circuit board. These are, of course, all machine soldered, and the specification of both solder and flux has to be very precise indeed if faults are to be kept to a minimum. Similarly, in the automobile field the cost of tin has risen so much in recent years that much more attention is now paid to the quality needed for such parts as radiators, body fillers and so on. The model engineer does not need a wide variety of soft solders, and many of the new developments have no relevance to our work. We can get over many and avoid most of the problems arising in machine soldering when we use manual methods. Nevertheless, we do benefit from the improved quality control compared with 'the old days' and especially from the reduction in impurities present – some of these could cause quite serious trouble.

Although nominally a tin-lead alloy, BS 219 does permit a small (¼% to ½%) amount of antimony in the normal soft solders, and a range of tin-lead-antimony alloys containing up to about 3% antimony is also included. The presence of antimony improves somewhat the strength of the cheaper solders having tin content less than the eutectic. There is some very slight increase in difficulty in wetting the parent metal with these under less than perfect conditions, so that more care is needed in surface preparation and fluxing. High antimony (above 0.6%) solders should not be used on high-zinc brass where the joint is highly stressed, or on galvanised sheet, as there is a suspicion that an antimony-zinc reaction may cause cracking, but for general joining work little trouble need be expected. Details of some of the solders listed in BS 219 are given in table 1 which appears opposite.

TABLE 1 – SELECTED TIN-LEAD BASED SOLDERS TO BS 219

(Sn=Tin; Sb=Antimony; Remainder is lead)

BS 219	Sn %	Sb %	Solidus °C	Liquidus °C	U.T.S. ton/in²	Shear† ton/in²	Remarks
A	63/64	0.6 max.*183	183	185	3.9	2.5	Very fluid, for close capillary joints and fine work.
K	59/60	0.5 max.*183	183	188	3.8	2.5	Fluid. A good 'all round' solder for electrical and mechanical use, easily worked. Good for brass and zinc.
F	49/50	0.5 max.	183	212	3.0	2.6	The usual general solder for sheet metal work: cheaper than Grade K. For all usual metals.
R	44/45	0.4 max.	183	224	2.8	2.3	General solders, cheaper than Grade F as the tin content is less, but higher liquidus.
G	39/40	0.4 max.	183	234	2.7	2.7	General solder, useful for brass
B	49/59	2½/3	185	204	3.9	2.7 ⎫	Antimony-bearing equivalents of 'F', 'R' and 'G'. Slightly stronger. *Not* for use on zinc, and not recommended for printed circuit soldering, or brass. "Plumber's Solder" *(see p. 28)*.
M	44/45	2.2/2.7	185	215	3.6	2.6 ⎬	
C	39/40	2.0/2.4	185	227	3.4	2.5 ⎭	
J	29/30	0.3 max.	183	244	3.0	2.5	Car body and similar surface filler. Can be used for joints to stand very low temperatures (<–60°C).
W	14/15	0.2 max.	183	276	—	—	
99C	99.5	0.5 TIN	227	228	—	—	Lead free, for jointing copper water pipes *(see p. 28)*.
T1/T3	99/99.9	—	227	MP 232°C	1.7	—	Useful for 'step soldering' and pre-tinning especially for bearing metals. Rather brittle if used alone.

NOTES: (a) *Low antimony grades (below 0.2%) are available, denoted by AP and KP for special applications.

(b) †As measured on a prescribed test joint on steel. UTS (tensile) figures are for solder 'as cast'.

(c) *Joint Gaps.* Whilst much larger gaps can be filled, maximum strength is achieved with gaps between 0.002 and 0.006 inch.

Extract from BS 219, reproduced by permission of the British Standards Institution.

BS 219 lists some 20 different alloys in this group, and others are available from various suppliers, some to US or DIN specifications and others to their own formula. Some of the latter are traditional and their trade names would be recognised by model engineers 100 years ago, whilst others are so new that they have not even been considered yet for inclusion in any standards. The choice, especially to the newcomer to soldering, can be bewildering.

Fortunately the needs of model engineers (and most small workshops) can be met by relatively few soft solders – at least, so far as general fabrication and jointing is concerned. (I shall be dealing with 'high temperature' work in a moment). Looking down the list it will be seen that the tin content and hence the cost diminishes as we go down the first part of the table. You will also notice that (e.g.) grade 'M' has the same tin content as 'R', but the addition of another 2% of antimony has given a strength equal to grade 'K'. As the tin content is reduced by 25% it will be very much cheaper, and in industry, provided that the antimony content does not matter, grade 'M' would reduce costs considerably. I find that for 95% of my soft soldering, both for fabrication and electrical work, grade 'K' – the '60/40 tin/lead' of the old textbooks – will serve. Compared with the value of the finished article there is NO point in changing to the 50/50 grade 'F' and even less in using any of the higher antimony grades. The amount used in the odd casual repair job where a 40% tin solder would be quite satisfactory is so small that it is just not worth the shelf-space to buy any, though I do have an odd piece of grade 'C' which I use occasionally when I want to build up a fillet.

On the other hand, it would be a wicked waste of tin to use a tinman's solder when filling a casting or a fabrication before painting. For this purpose I keep a stick of 15% tin, grade 'W'. This may be a relic of the days when I lived in a house with lead pipes and 'wiped' joints, but it has come in handy over the years filling dents in the car or depressions in a fabrication where I have been too fierce with the beating-hammer. The only other material from this list which I use is 'pure' tin – actually 99.75% pure. This I use for two purposes. First, for pre-tinning steel bearing shells (both full-size and model) before white-metalling them. This is the only way to ensure a perfect bond between the white metal and the shell. (White metal, or 'Babbitt', is a tin-bearing alloy and *should* bond to steel but under the hammering of the shaft in the bearing it does tend to craze if the tin coat is not laid on first). The second use is when the really difficult bonding operation turns up – usually the repair of some ancient metal article – when I find that tinning the surfaces with pure metal first does improve the capillary flow of the proper filler into the joint .

SPECIAL SOFT SOLDERS

(a) **For high temperatures** There are several grades of solder which will withstand higher working temperatures than the simple tin-lead series. Here it should be noted that more important than the loss of strength of tin-lead solders above 100°C is their tendency to 'creep' – failure occurring after some considerable time under stress. Most 'ordinary' solders lose about 75% of their room temperature tensile and 35% of their shear stress at 100°C and at 170°C (boiler steam at 100 lb/sq.in) the

TABLE 2 – COMMON 'HIGH TEMPERATURE' SOFT SOLDERS

BS No.	% Tin	% Lead	% Antimony	% Silver	Solidus °C	Liquidus °C	UTS at 100°C lb/sq.in
95A	95	—	5	—	236	243	2900
96S	96.5	—	—	3.5	M.P. 221		
5S	5	93.5	—	1.5	290	300	2750

tensile strength is less than one tenth of that at room temperature. This is for bulk solder; the loss in strength of the actual joint may not be quite so pronounced. Nevertheless, there is little to be said for using normal soft solders where there is a *stressed* duty at temperatures above 100°C. The common 'High Temperature' solders are as in Table 2.

For comparison, the 100°C ultimate tensile strength of 60/40 solder BS grade K is about 600 lb/sq.in and at 150°C the *safe life* at a stress of 100 lb/sq.in is only about 10 hours.

There are several proprietary brands of 'high temperature' soft solders in common use by model engineers. 'Comsol' is identical to BS grade 5S above, as is 'LST5'. HT3 accords with BS grade 95A. LM5 is a silver-cadmium alloy with the very high melting range of 338-390°C, and returning an 'as cast' strength of 2700 lb/sq.in at 200°C. It is a little difficult to work (clearly an ordinary 65 watt 'soldering iron' will not serve) but provided that both surfaces of the joint are 'tinned' with the alloy reasonable capillary jointing can be achieved. LM15 is a cadmium-zinc alloy containing 5% of silver, melting range 280-320°C. It has a very high strength (about 11 tons/sq.in at room temperature) but is rather brittle. These solders are all useful in circumstances where high temperature brazing is not possible, provided that the strength factor is borne in mind. It is, however, questionable whether it is really necessary to consider 'strength' when the material is used only to caulk joints which rely on (e.g.) rivets for their strength. Provided the solidus lies well above the operating temperature BS grade 5S should serve.

(b) **Low melting-point solders** These are not often needed in model or light engineering but it is useful to know that they exist. BS 219 Grade 'T' is a 50% tin, 18% cadmium, 32% lead alloy with a melting point of 145°C – there is no melting range. It is intended chiefly for making soldered joints close to insulating materials which might be damaged by the temperature of normal solder, but obviously can be used in step soldering as well. The next step down is found with a 50/50 tin-indium alloy with a melting range of 117-125°C to the German specification DIN 1707. Other alloys are based on bismuth, a 26% tin, 20% cadmium, 54% bismuth alloy having a melting point of 103°C and, of course, the well known 'Wood's Metal', 15.4% cadmium, 30.8% lead, 15.4% tin, 38.4% bismuth, with a melt-point of only 71°C. Though better known for its use in the manufacture of 'joke' teaspoons it can be used as a solder if need be!

(c) **Solder Paste** This is a special *form* of solder rather than a special type. The material is simply finely divided solder 'dust' mixed with flux to make a paste or cream. This can be spread over a surface or disposed within a joint and then gently heated externally up to the soldering temperature. Most manufacturers provide a range of solders in this form, each with a range of fluxes; pure tin paste is available for the re-tinning of cooking vessels or coating bearing shells, and both 'neutral' and highly active fluxes can be had if need be. The 'Standard' grade put up by most makers, and that supplied when you ask for 'Solder Paste' with no speci-fication, will be BS Grade 'C', 40% tin, melting range 185 to 227°C, incor-porated into an 'active' flux which must be washed off after the joint is made. The solder contains 2-2.4% antimony, so that the paste is unsuitable for use on galvanised material, but it is a good all-round paste which has a multitude of uses. However, the following solders CAN be had if need be: BS Grade 'K', Pure tin, BS 95A High temperature solder (melting range 236-243°C) and a special grade for use on stainless steel. These can all be supplied with the 'ordi-nary' flux, or with non-corrosive or even 'protective' fluxes, and each manufac-turer has one or more special formula-tions to meet particular needs. The main problem facing the small user is that of 'supply'. Most tool dealers stock the standard grade, but 'minimum order' conditions may apply if anything out of the ordinary is asked for. I find that the standard material meets almost all my needs and the only 'special' I have needed is pure tin. While there are occasions when an out-of-the-ordinary grade might have helped over the last 40 years or so I have always managed without in the end!

(d) **Aluminium Solders** The soldering of aluminium presents two problems. First, the high thermal conductivity of the metal means that heat can 'run away' from the joint area very quickly; heating methods suitable for normal soldering may be inadequate. Second, the oxide film on aluminium forms very quickly indeed and is highly resistant to chemical attack. We shall see what ef-fect these have when we come to look at 'techniques', but these two factors have made the soldering of aluminium a matter of considerable difficulty in the past. Recent research has, however, re-sulted in the formulation of solders and fluxes which make matters rather easier, and the soldering of pure aluminium sheet is now reasonably easy. If any aluminium *alloy* is to be soldered, how-ever, it is best to consult the manufac-turers either of the metal or of the solder.

TABLE 3 – ALUMINIUM SOLDERS
(Courtesy Fry's Metals, Ltd.)

Grade	Solidus °C	Liquidus °C	Remarks
ALCA 'P'	170	220	Used with flux for jointing cables. Cadmium free.
ALCA 'Z'	170	179	For hand soldering with flux on pure aluminium.
FRYAL	177	312	For 'Abrasion' techniques. Has a long plastic range.
'LM'	199	205	Tin-zinc alloy, normally used without flux.
703 Alloy	199	310	For tinning aluminium before soldering with tin-lead solders. Intended for cable jointing.

Fry's Metals Ltd list those in Table 3, most of which are tin-zinc alloys with a cadmium addition. The 'Fryal' grade has a special attraction for model engineers as it can be used rather like plumber's solder to 'build up' aluminium castings, either to rectify defects or to correct machining errors.

(e) **'Universal' Solders** So-called Universal solders are on the market, the advantages claimed being that they will bond with almost all metals, thus obviating the need for a range of alloys. The analysis of such metals is usually confidential, so that it is necessary to obtain a sample and make your own tests. I have used these, generally with success, but find no advantage over the ordinary engineering alloys except in the case of repairs to objects made of unknown metals. Here there is an advantage; the melting point is usually low (typically 210°C) and though this is too high for pewter it does serve with many die-casting alloys. They are not true 'capillary' solders; the technique is to 'tin' the surface of both parts, using a flux except on aluminium, magnesium, or zinc, and then to 'sweat' the two

parts together. It is perhaps worth having a little in the workshop, for the model engineer is usually credited with the ability of being able to 'mend anything' and provided that you make strength and solderability tests on pieces of scrap to satisfy yourself as to quality (and to master the technique) you may well do great service to your neighbours. I would not recommend these alloys as a substitute for BS grade solders in any serious fabrication work in the materials we normally use; apart from anything else the cost is rather more than 'normal' solders.

Availability of Soft Solder Fig. 16 shows the commoner forms in which solder is supplied. At the top is the remains of a 1-lb slab of plumber's solder, while just below is a ½-lb stick (or bar) of tinman's solder. This is still the most convenient form, and at one time the 'slab' and 'stick' were the only ones. On the left is a coil of plain solder wire, 18 gauge, though it can be had in sizes up to ⅛in (3mm) in most grades. The 'High Temperature' solders are usually sold in wire form. In the centre you see two sizes of 'preforms', home-

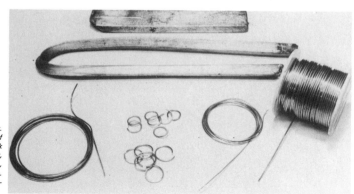

Fig 16 *Forms of soft solder. Top, part of a bar or ingot of 'Plumber's' solder, with a stick of 'Tinman's' solder below. Left, Tinman's solder in wire form, with some 'preforms' alongside. Right, two thicknesses of flux-cored solder wire.*

made by winding the wire on a mandrel and cutting lengthways, while at the right is a drum of 18 gauge and a coil of 24 gauge multi-cored electronic solder. There ARE other forms — pellets, foil, shaped preforms etc, but these are for industrial use — you could get a 56-lb ingot if you liked!

The main problem these days may well be supply. The 'We are never asked for that' syndrome is now endemic in this country and it is a sad fact that few ironmonger's assistants know anything about solder at all. Engineer's tool dealers are better, but even here the stock is likely to be limited to the needs of the local area. However, most of the latter, and the model engineer's suppliers, can obtain any grade against a firm order, and in the last analysis you can write to the actual manufacturers asking for the address of a sales outlet ('shop' to you) which stocks that particular grade and form. But don't be put off; seek out the grade you need in the form which is most convenient to you, and get enough to last for a year or so; it won't go bad in storage!

Since this chapter was written new regulations relating to the use of solder in water systems have come into force. The use of lead or lead alloy is now forbidden, and LEAD-FREE solder is now obligatory for (e.g) jointing copper pipes, the seams of or repairs to cylinders, etc. The appropriate alloy is BS. 219 Grade 99C. This is a copper-tin eutectic (0.5% copper/99.5% tin) with a melting point of 228°C — there is no 'plastic range'. It is available both as stick or wire from plumber's merchants or D.I.Y. stores.

"NOMENCLATURE" is most important here. True "Plumber's solder" (BS Grade D or W) has a long pasty stage when solidifying, as it was used when "wiping" joints on lead pipes. The "solder that plumbers use" on copper water pipes (BS 99C etc) solidifies almost instantaneously. It is always wise to specify solder by referring to the British Standard 219 grade letter or number — Tables 1 and 2.

Chapter 4

Brazing Alloys

The basic brazing alloy is BRASS, often known as 'Spelter', which is an alloy of copper and zinc. The copper-zinc system does not show any 'eutectic' in the melting-point diagram, but the presence of other constituents does cause a short melting range. These copper/zinc alloys have fallen out of use in industry because the silver-brazing alloys offer equal strength at much lower brazing temperatures, and though the alloy itself is more expensive the heating costs are very much less. Nevertheless, 'common spelter' is not to be disdained for work on steel or copper as, for the model engineer, the cost of the brazing rod is often more important than that of gas! There is the added point that brass wire is often available in the workshop anyway (or even brass turnings) and this will braze quite satisfactorily provided the brass is not the lead-bearing free-cutting variety ('screw brass'). Fig. 17 shows the end view of a joint I have made using 60-40 brass spelter, and you can see that there has been good penetration to the joint – the rod was applied to the toe of the steel angle. This was made with no special preparation, nor with any 'scrabbling about' with pointed wire to get the metal to flow. (I shall have more to say about this when we come to fluxes). True, the workpiece was really hot, but I think that the great master 'LBSC' was exaggerating a bit when he told his readers to 'get it so hot that you could almost see through it'! The temperature in this case was around 900°C. On another point I cannot refrain from showing you how my grandfather used to

Fig. 17 *Sectioned view of joint brazed with 60/40 'common' brass wire, showing the excellent penetration.*

Fig. 18 *Brass swarf melted into impressions from letter-stamps. 'Sifbronze' flux was used.*

mark his tools in his apprentice days. Fig. 18 shows a piece of steel flat which has been roughly stamped with $^{7}/_{32}$-inch letter punches and then filed smooth. The piece has been coated with a mixture of brass swarf (just 'any old stuff' taken from my swarf bin) and wet flux, and then heated until the brass melted. It has then been filed back to the steel again, and polished. I have heat blued it to make the letters stand out more. Quite a handy way of marking things and making nameplates if you have a steady hand!

BS 1845 lists five 'copper-zinc' brazing alloys. (Note that this specification was revised as from May 1984; the old CZ3 alloy, a straight 60/40 brass, no longer appears). The following table, reproduced by permission of the British Standards Institution, gives the details, but omits the limits on the impurities and the lead content, which must be below 0.02% in all cases.

The strength of these alloys approximates to that of cast brass, but as in the case of all capillary jointing the JOINT strength may be higher. The nickel-bearing alloy, CZ8, is white in colour and can be used on (e.g.) steels and nickel silver to avoid a 'colour clash'. The alloy most likely to be found 'on the shelf' is CZ6, but older brazing spelters may well be straight 60/40 copper zinc alloy containing traces of tin and lead, and with a melting range of 885-890°C.

Silver-brazing alloys The high melting-point of the conventional copper-zinc alloys led to a search for equally strong alternatives, and as long ago as 1846 Holtzapffel described a 'silver solder' made by alloying 2 parts of silver to 1 part of common brass. However, contrary to common supposition, today's silver brazing alloys are based on the **copper-silver** eutectic system, a simplified diagram of which appears in Fig. 19. You will see that it is of similar shape to the tin-lead system of Fig. 9, but the temperatures are much higher, the eutectic lying at 780°C. By adding zinc to this alloy one of two effects can be achieved: either the melting range for a given silver content can be re-

TABLE 4 – COPPER-ZINC BRAZING ALLOYS

BS 1845	Cu. %	Zn %	Tin %	Si %	Mn %	Ni %	Solidus °C	Liquidus °C
CZ6	58½-61½	Rem	⟨0.2	0.2-0.4	—	—	875	895
CZ6A	58½-61½	Rem	0.2-0.5	0.2-0.4	—	—	875	895
CZ7	58½-61½	Rem	⟨0.2	0.15-0.4	0.05-0.25	—	870	900
CZ7A	58½-61½	Rem	0.2-0.5	0.15-0.4	0.05-0.25	—	870	900
CZ8	46-50	Rem	⟨0.2	0.15-0.4	⟨0.2	8-11	920	980

In all cases the lead content is not to exceed 0.02% and the iron content must be below 0.25%.
(Reproduced with permission from the British Standards Institution.)
Cu=Copper; Zu=Zinc; Si=Silicon; Mn=Manganese; Ni=Nickel.

duced, or we can achieve the same melting range but use less silver. This tripartite (or 'ternary') system has a eutectic at about 677°C — over 200°C lower than the solidus of the older brazing alloys.

The addition of cadmium to this ternary system, first suggested about 50 years ago, resulted in even lower solidus temperatures without materially changing the melting range and with little loss of strength. Though the eutectic of an alloy of four elements is hardly identifiable, solidus temperatures as low as 600°C were achieved with melting ranges very short indeed — 10-20°C. These were the 'easy-flowing' alloys which had such a marked effect on the pre-war model engineering field. The much lower working temperatures,

better fluidity (partly due to more highly developed fluxes) and reduced cost compared with the 'straight' silver solders not only removed almost all the difficulties then met in brazing, but (perhaps more important to some!) permitted the fabrication of complex machine parts which otherwise would have had to be 'chewed out of the solid' or would have needed the making of patterns and castings.

More recently the effects of small additions of tin have been studied and resulted in the introduction of a range of cadmium-free brazing alloys which have very similar characteristics, though perhaps not quite so pronounced. This system can now offer solidus temperatures as low as 630°C with a melting range of no more than 30°C, and with

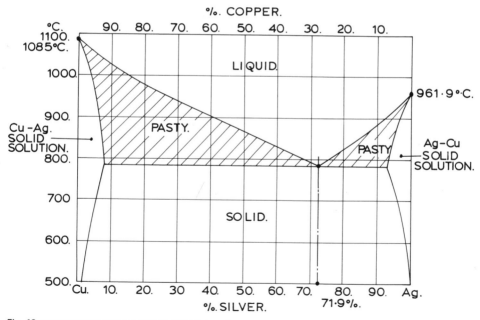

Fig. 19 *Simplified Melting-range diagram for Silver-Copper alloys.*

TABLE 5 – THE BS 1845 RANGE OF SILVER-BRAZING ALLOYS, WITH SOME EQUIVALENT 'TRADE NAMES'

BS 1845 No.	Ag %	Cu %	Zn %	Cd %	Sn %	Mn %	Ni %	Solidus °C	Liquidus °C	Fry's Metals Ltd.	Johnson Matthey Ltd.	Thessco Ltd.	Remarks
AG1	50	15	16	19	—	—	—	620	640	FSB No. 3	Easyflo No. 1	MX20	
AG2	42	17	16	25	—	—	—	610	620	FSB No. 2	Easyflo No. 2	MX12	
AG3	38	20	22	20	—	—	—	605	650	FSB No. 1	Argoflo	AG3	
AG11	34	25	20	21	—	—	—	610	670	FSB No. 15	Mattibraze 34	MX4	
AG12	30	28	21	21	—	—	—	600	690	FSB No. 16	Argoswift	MX.O	
AG9	50	15½	15½	16	—	—	3	635	655	FSB No. 19	Easyflo No. 3	MX20N	Nickel-bearing
Cadmium Free Alloys													
AG14	55	21	22	—	2	—	—	630	660	FSB No. 29	Silverflo 55	M25T	
AG20	40	30	28	—	2	—	—	650	710	—	Silverflo 40	M10T	Tin bearing
AG21	30	36	32	—	2	—	—	665	755	FSB No. 33	Silverflo 302	MOT	
AG13	60	26	14	—	—	—	—	695	730	FSB No. 4	Silverflo 60	HO	
AG5	43	37	20	—	—	—	—	690	770	FSB No.5	Silverflo 43		
AG7	72	28	—	—	—	—	—	M.P. 780		FSB No. 17	Agcu Eutectic	H12	Silver/copper eutectic
AG18	49	16	23	—	—	7.5	4.5	680	705	FSB No. 37	Argobraze 49H	M19MN	Nickel-manganese
AG19	85	—	—	—	—	15	15	960	970	—	15 Mn-Ag	—	Silver-manganese
From BS 1845/1977													
AG10	40	19	21	20	—	—	—	595	630	FSB No. 10	DIN Argoflo	MX10/DIN †	
AG15	44	30	26	—	—	—	—	675	735	FSB No. 39	Silverflo 44	M14 †	
AG16	30	38	32	—	—	—	—	680	770	FSB No. 25	Silverflo 30	MO	
AG17	25	41	34	—	—	—	—	700	800	FSB No. 23	Silverflo 25	L18 †	
From BS 1845/1966													
AG4	61	29½	11	—	—	—	—	690	735	—	—	H1	

(Solidus and Liquidus columns are Approximate, in °C.)

†German DIN specification applies.

(Reproduced by permission of the British Standard Institution.)

In using this table it must be remembered that though the 'Commercial Types' listed above do conform to BS 1845 in the grades indicated, each manufacturer will have developed the metallurgy of his own products. Slight differences of alloy content, still within the limits of the British Standard, will result in small differences in the melting range. These differences are seldom significant in service.

strengths equal to or very near those of the cadmium bearing series. Thus the model engineer now has available straight silver-copper-zinc, silver-copper-zinc-tin, and silver-copper-zinc-cadmium alloys, together with a few containing also nickel or manganese; a considerable change from the days when all brazing had to be done with 'spelter'!

There has been a fair amount of discussion on the hazards of cadmium-bearing alloys recently, and I deal with this in some detail later in the book. In general, the risk need not be serious for the model engineer or small user even if he does a fair amount of brazing, provided that he is 'sensible'. My own experience is that sheer discomfort enforces enough ventilation to reduce the concentration of hazardous fumes (both from flux and alloy) simply because otherwise I get too hot! However, there are now these cadmium-free alloys, some of which are quite as easy-flowing, and the additional cost is not great.

I am faced with some difficulty in listing the silver-brazing alloys. BS 1845 has just been revised, and some alloys in the 1977 standard have been left out. Some well-known 'brand names' are formulated to DIN or ASME standards for which there is no direct British equivalent. Again, you may find that two alloys, both conforming to BS 1845 but by different manufacturers, may be shown in their lists as having different liquidus/solidus points. This last is to be expected, of course, as all Standards permit a *tolerance* on the alloy content. For example, 49Ag/14Cu/18Zn/19Cd and 50Ag/15Cu/17Zn/18Cd both fall within the limits of alloy AG1, but the melting range will be slightly different. Finally, almost all manufacturers have

developed alloys for special purposes which have considerable advantages, but which conform to no national standard. I am afraid that the basic purpose of these standards is not always understood. The objective is not to identify the IDEAL material for ALL purposes, but to set up specifications such that for the majority of (in this case) brazing operations users can quote a Standard Number and be sure that the material is, within limits, the same no matter who supplies the alloy.

What I have done is, first, to draw up a table based on BS 1845/1984 listing the alloys it contains, but to give the mid-range alloy content only. This means that where BS permits from 41% to 43% silver I have tabulated it at 42% — which is the 'nominal' content. I have also included the three alloys which appeared in BS 1845/1977, and one or two from earlier issues of the Standard, where they are still available from manufacturers. So, the first part of Table 5 covers the current revision, and the remainder shows alloys which may still be available to earlier revisions. Second, I have included in Appendix I a set of tables summarising the *manufacturer's* data. Note that these tables are for conventional silver-bearing brazing alloys; the phosphorus or so called 'fluxless' alloys are dealt with later.

It must be confessed that Table 5 presents an almost bewildering range of choice! Some of the alloys are, of course, for special applications. AG7, for example, is for joints under high vacuum, and the manganese-bearing alloys have their main use in the brazing of carbide tips to cutting tools. The non-industrial user will probably be more concerned with the working temperatures than with the actual constituents and perhaps it may be as well to say a little about this before commenting on the alloys individually. But please bear in mind that there are exceptions to all generalisations!

BS 1845 alloys are all 'capillary brazing' materials, but some are more capillary than others, and as a result, some produce more pronounced fillets at the joint. Look at Fig. 20. This is a type of joint which needs a fillet. The joint width at 'b' is almost three times that at 'a' so that the stress caused by the force on the upright is reduced. However, in Fig. 21 we have a strapped butt-joint, and here a fillet is unnecessary. We need, on the other hand, good penetration of the joint gap. In general, an alloy with a long melting range will tend to form fillets, and one with a short range will tend to have better penetrating power (see Fig. 22). So, other things being equal we select from Table 5 accordingly. The second point is the actual solidus temperature. Fig. 23 shows part of a model which is made up of something like 30 pieces (not counting the butt-straps on some joints) and although it was made in three sections, and a number of joints were brazed at the same time, it was clearly necessary to join some parts to others which already had brazed joints. This requires us to use higher melting-point

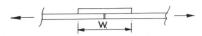

Fig. 21 *A fillet is not needed here, as the alloy is in 'shear'.*

alloys for the first steps, turning to lower and lower figures as the fabrication proceeds. Clearly we need alloys with quite a wide difference in their solidus temperature for this sort of fabrication, and these can be selected from Table 5. (The technique of 'step' or 'sequence' brazing will be dealt with in more detail later, but it may be worth mentioning at this stage that it is not *always* necessary that the solidus of the alloy used at one stage to be above the liquidus of the alloy used in the next.)

The final criterion to be considered is the *joint gap*. As I have already pointed out, there must *be* a gap if the alloy is to flow (without this the joint must depend entirely on fillets) but there are cases where the gap *has* to be larger than might be desirable. Some alloys will penetrate a radial gap of 0.001 inch (0.025mm) but will not enter one of more than 0.004 inch, while others which will fill a gap as much as 0.008 inch cannot penetrate anything below 0.003 inch wide. (Obviously wide gaps are to be avoided if possible, if only because it costs more to fill them!)

Thus the final selection of the alloy to be used may well have to be a compromise between these various requirements; it is seldom that one will meet all, and rare for there to be none that will serve. (It goes without saying, of course, that the alloy must be compatible with the base metal to be joined; you would not expect to be able to use AG19 on common brass!) Let us now look at some of the alloys in more detail.

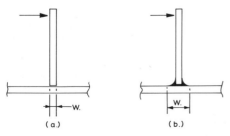

Fig. 20 *The stress at the joint 'a' will be far higher than that at 'b', where the effective width 'W' is increased by the fillets.*

Fig. 22 *Fillets. Left, alloy with 10°C melting range. Centre, alloy with 30°C melting range. Right, alloy with 80°C melting range.*

AG1 & AG2 These are the original 'Easy Flowing' cadmium-bearing silver alloys, and are the ideal general purpose alloys. They can be used on most metals (if the right flux is applied) including stainless steel, some carbides, and on the aluminium bronzes provided that the aluminium content is less than 2%. There is little difference in the working temperatures, but AG1 has slightly better capillary action and forms neater fillets. Its corrosion resistance is slightly better and the ductility is a little higher than AG2. However, AG2 is the cheaper alloy and can be recommended for all general purpose work, including boiler-making. Joint Gap: AG1 0.001-0.004 inch; AG2 0.002-0.004 inch.

AG3 This alloy has good capillarity, though not quite as penetrating as AG1 and AG2. It forms small to moderate fillets. The liquidus is low, yet the rather larger melting range does not present serious problems with liquation if care is taken. The alloy can tolerate a rather wider joint gap than AG1 and AG2. It is cheaper than those with higher silver content. Recommended for butt joints (Fig. 19) and where larger or irregular joint gaps may be present. Joint Gap 0.002-0.006 inch.

AG11 & AG12 Alloys of progressively lower silver content and hence cost.

Although the solidus differs little from the higher silver alloys the melting range is larger and they can be used for step brazing. Care must be taken to avoid liquation, especially with AG12.

Fig. 23 *An example of 'step' or 'sequence' brazing. This fabrication contains more than 30 pieces.*

Wider joint gaps are needed. Joint Gap: AG11 0.003-0.008 inch; AG12 0.004-0.008 inch.

AG9 A nickel-bearing alloy with a short melting range. Chiefly used for brazing tungsten carbide tool tips. However, the alloy also has an application where corrosion from sea-water is likely. Capillarity is poor and the alloy is very sluggish, so that normally it is applied as a preset foil between the mating surfaces. Substantial fillets can be formed with conventional alloy application.

AG4 Cadmium free. Though deleted from BS 1845 in 1977 this, one of the high-silver alloys, is still available, especially for brazing electrical conductors. It is white in colour and resists oxidation in service. Joint Gap 0.002-0.006 inch.

AG13 Very similar in content to AG4 and with a smaller melting range. Its low zinc content recommends it for brazing nickel and nickel alloys. Joint Gap 0.002-0.006 inch.

AG14 The cadmium-free tin-bearing substitute for AG1 and AG2. Fluidity is very good and deep joints can be filled, with small, neat fillets. An excellent all-round brazing alloy which can be used on the same base metals as AG1 and AG2. This alloy can be recommended for those practitioners who wish to avoid the use of cadmium. Joint Gap 0.002-0.005 inch.

AG20 & AG21 The low silver alloys of the tin-bearing range which starts with AG14. Cheaper, but with wider melting ranges, and hence needing more care to avoid liquation. Can be used in sequential (step) brazing. Joint Gap 0.002-0.008 inch.

AG5 Less expensive than AG13 and AG14 but with a much wider melting range and higher solidus. The latter makes it useful for the first stage in multi-step brazing provided care is taken to avoid liquation. It has been recommended for use with nickel-base alloys where stress-corrosion cracking is anticipated. Joint Gap 0.003-0.008 inch.

AG7 Of little use to the model engineer, this silver-copper eutectic is used chiefly for the brazing of components subject to high vacuum, employing controlled atmosphere furnace brazing techniques. A very fluid alloy, penetrating extremely small joints.

AG18 A manganese/nickel bearing alloy, formulated to give enhanced wetting and improved shear strength for brazing carbide tool-bits. The flow characteristics are not too good, and preset foil is advisable if the joint area is of any size.

AG19 The silver/manganese eutectic, included for completeness.

The following four alloys do not appear in the current (1984) revision of BS 1845, but form part of a useful series of alloys for step brazing. All are commercially available and most can be released against DIN specifications.

AG10 The silver content falls midway between AG2 and AG3, and it has the lowest solidus of any of the silver brazing alloys, though the liquidus is 10°C higher than AG2. The fluidity is good, but the alloy forms neat fillets also. Cheaper than AG1 or AG2, but with slightly longer melting range. Joint Gap 0.002-0.006 inch.

AG15, AG16 & AG17 These provide a range of cadmium-free alloys of progressively lower silver content and in conjunction with other alloys provide reasonable steps in operating temperature for sequential brazing. Liquation can be a problem with AG16 and AG17 if the heating arrangements are not good. With the other cadmium-free alloys a range of silver content from 60% down to 24% is available. Joint Gap: AG15 & AG 16 0.002-0.006 inch; AG17 0.003-0.010 inch.

As already suggested, these British Standard alloys do not represent the full spectrum of alloys which are available, and when those to DIN or U.S. specifications are added the list becomes very long. For the model engineer who uses 'silver solder' only for the occasional fabrication of fittings or small components then AG2 or the cadmium-free alternative tin-bearing AG14 will cover almost all requirements. Both are easy to use and require but moderate temperatures. I tend to use AG1 for joints which include a riveted butt-strap (Fig. 21) just in case the riveting has closed up the gap to very close limits (it has better penetrating power) and when the final appearance of the joint is of paramount importance, but this is, perhaps, counsel of perfection! It may be that I was 'brought up' on AG1! For more elaborate fabrications, where sequence brazing is needed the series AG17/AG14/AG2 gives a respectable overlap, or Ag14/AG2 if only two steps are needed. My own 'armoury' also includes alloys of 16% and 24% which, along with AG14, AG1 and AG2 does permit multiple step brazing when needed.

Phosphorus-bearing alloys (so-called 'self-fluxing') These alloys are often referred to as 'fluxless', which is quite wrong. Intended for use on copper, the inevitable oxides which form at the working temperature are mostly dissolved in the alloy. The phosphorus content then reacts with the oxides to form a copper phosphate which floats as a slag on the surface of the braze. This slag then protects the molten alloy from further oxidation. (The slag is harmless and there is no metallurgical reason for removing it.) These alloys are cheaper than the conventional silver-brazing materials and for large workpieces requiring a considerable amount of brazing rod they have found increasing favour in recent years, despite certain disadvantages.

They were devised specifically for brazing COPPER, in situations where subsequent removal of flux might be difficult, and though they can be used on some copper-based alloys of appropriate melting-points they must **NOT** be used on iron or steel, **NOR** on any nickel or cobalt based alloy, or on aluminium based material. Indeed, the only materials quoted by BS 1723 as being 'possibly' suitable for these brazing alloys other than copper and copper alloys are tungsten and molybdenum. **This is important,** as if an attempt is made to braze other than 'permitted base metals' the alloy will *apparently* perform quite well, with good flow and penetration. On test, however, the joint will be found to be very brittle indeed. The model engineer and jobbing brazier would do well to confine any use of phosphorus-bearing brazing alloy to pure copper.

There is one reservation entered by all brazing alloy manufacturers which is relevant; they warn that the finished joints should *not be exposed to '. . . hot, sulphur-containing gases, or* where the service environment is oxidising at temperatures above about 200°C'. This rubric indicates that the material is *not* suitable for use in coal fired boilers, and a report by Mr. P. D. Wardle in *Model Engineer* (1 Feb. 80, p. 143 et seq) gives details of a boiler failure where the braze was made with an alloy corresponding to BS 1845, Grade CP3. Analysis of the joints showed severe corrosion from sulphur. True, this was not a silver-bearing alloy, but metallurgists suggest that no more than marginal improvement will follow the use of silver-phosphorus material. The reduction in alloy costs, and the marginal convenience of requiring no additional flux is negligible compared with the cost and value of a model boiler, and these alloys are best avoided for this application. It should be noted that though these alloys have a high tensile strength (30-40 tons/sq.in) they are considerably less ductile than copper, with elongations in the region of 10% or so. CP2 is rather more brittle still.

In their proper place they have their uses, and being zinc-free will not suffer from de-zincification in water systems, whilst their similarity to copper in the electro-potential series makes them suitable for brazed electrical joints. The table below (Table 6) gives the particulars — again with the 'midrange' proportions of the constituents. It will be noted that a 'working temperature' is given which lies below the liquidus. This is possible because at this temperature the phosphorus-bearing alloys are around 95% liquid and capillary flow can take place despite the presence of some unmelted material.

Aluminium Brazing Alloys Although aluminium is 'brazeable' it does present difficulties, especially for those who use the process but rarely. The liquidus of most filler materials is only about 70°C below the melting-point of pure aluminium and the risk of melting the parent metal is high. (Though a coating of kitchen soap can give warning, as it turns brown at about the brazing temperature.) The heat conductivity of aluminium is high, and this, again, incurs the risk that the operator may use too HOT a flame in order to speed up the heating of the work. The occasions when the model engineer actually

TABLE 6 – COPPER-PHOSPHORUS BRAZING ALLOYS

| BS 1845 Grade | Phos. % | Silver % | Copper % | Ant. % | Approximate Temperature | | |
					Solidus °C	Liquidus °C	Working °C
CP1	4.6	14.5	Rem.	—	645	800	800
CP2	6.5	2.0	Rem.	—	645	825	740
CP3	7.4	—	Rem.	—	710	810	730
CP4	6.0	4.0	Rem.	—	645	815	710
CP5	6.0	—	Rem.	2.0	690	825	740
CP6	5.2	—	Rem.	—	710	890	770

In all cases the cadmium content shall not exceed 0.025% and total of all impurities 0.25%.
CP5 and CP6 are newly introduced to BS 1845 in 1984.
The correspondence between these alloys and a few well-known brand names will be found in Appendix I.
(Reproduced by permission of the British Standards Institution.)

needs to braze aluminium are few. For most applications low temperature soldering will suffice, and as the metal is easily welded – which needs about the same skill as brazing in this case – we need not spend much time on it.

The alloys are almost all silicon-aluminium alloys though one, BS 1845 4145A (formerly known as AL1) contains some copper, and a few include a small proportion – 1% to 2% – of magnesium. For model engineering use only two need be considered, and these are both drawn from the former 1977 British Standard.

AL1 (4145A) 9/11% Si, ‹0.6% Fe, 3/5% Cu, ‹0.15% Mn & Ti, Al, remainder. Range 520-585°C.

AL2 (4047A) 11/13% Si, ‹0.6% Fe, ‹0.3 Cu, ‹0.15% Mn & Ti, Al, remainder. Range 575-585°C.

For those needing to braze aluminium (or its alloys) as a matter of importance the advice and data sheets of alloy manufacturers should be sought, both as to the most suitable alloy and flux, and procedure.

Other Brazing Alloys It might be thought that the previous pages list sufficient brazing materials for any purpose. By no means – though fortunately they WILL cover all the foreseeable requirements for the making of models! But BS 1845 lists no less than 13 alloys based on nickel, more than a dozen palladium-bearing and, of course, the gold-brazing alloys as well. (The use of which is far from being confined to the jewellery trade; they are extensively used where corrosion resistance is important). None of these need really concern us, but it's nice to know that they are there!

Availability of Brazing Alloys 'Spelter' or copper-zinc alloys were at one time universally supplied in the form of granules, in various degrees of coarseness, but I have not seen any of these for many years. My last consignment of true 'spelter' came in wire form, $\frac{1}{16}$ inch dia., in a half-pound coil.

The silver-brazing alloys were orinally manufactured in sheet form and sold in sheared strips of various widths and thicknesses. These are still available though not very common. I have a few pieces of various alloys in this form which I have kept for use when I need to place small snippets of alloy rather than use a rod. But the most usual form these days is in hard-drawn rods, from 1mm dia. up to 5mm, of which perhaps the 1½mm and 2mm sizes are the most used. These rods are from 550mm to 600mm long – rather shorter than the usual welding rod, but easier to manipulate as a result! Below (and including) 1mm dia. almost all grades can be had as wire coils, down to 0.5mm (about 26 S.W.G.) and I commend this form to you – it is very convenient for making preforms (see Fig. 16) and for small jobs where the 1½mm rod would be too heavy. I keep a small coil of both ¾mm and ½mm wire on hand in AG2.

Unfortunately these alloys are expensive, and the minimum order is (with most manufacturers) 0.5kg, so that unusual forms, and the very useful 'foil' – made down to 0.08mm thick – is not usually stocked by tool or model supply dealers. Commercial preformed rings are not worth buying unless you need a large number, as it is more economic to wind them up from wire. As in the case of soft solder, if you run into difficulties over supply an enquiry to the manufacturers will usually bring an indication of where 'small quantities' can be had.

A few of the alloys are available in the 'flux-coated' form, and these are appearing in DIY shops and motor accessory dealers. The Johnson Matthey 'Mattipak', containing 5 rods, is now offered with nine alternative grades of alloy, both cadmium-bearing and cadmium-free. However, you must remember that the minimum order by the retailer may well be 1kg, so don't expect him to order you a single pack! This matter is complicated by the fact that the price of silver varies from day to day, and invoices are usually on the 'price ruling at date of delivery' basis.

Finally, AG1 may be had as a powder mixed with flux – 'Brazing Paint' – suspended in an organic liquid. This is extremely useful for very delicate work, but does tend to 'go hard in the pot' if kept for any great length of time. I cover this material in more detail on page 75.

Conclusion The non-industrial user can be well served with very few brazing alloys and, for those who do very little, types AG2 or the cadmium-free AG14 will be found by far the easiest to use. For others, the temptation to seek ever-lower silver content alloys in order to save costs should, in my view, be resisted. Unless you are using the stuff by the kilogram the cost of brazing alloy is a very small part of the cost of other materials or of the 'value' of the finished product. It is far better to seek the material which is the easiest to use and which gives the most satisfactory joint *when made in your workshop*. In general, this means the use of alloys with short temperature ranges unless the design needs a pronounced fillet.

If step or sequential brazing is needed, then it may be necessary to use some of the wider-range alloys and to take care to avoid the problems of liquation – which, though they can be daunting when first met, do diminish with more experience.

For 'emergency use' the old (indeed ancient) process of spelter-brazing can be used; it is effective, strong, and cheap, and the absence of any 'proper' brazing spelter need not bother you, for 60/40 brass wire will serve. This needs higher temperatures but they are quite within the reach even of a paraffin blowlamp on small jobs. For those who may disdain such procedure it is, perhaps, worth mentioning that until relatively recently all copper steam pipes in the navy were brazed to their flanges using 'spelter' and even on the Great Western Railway locomotive dome-covers were so brazed. (In passing, it is NOT true that the GWR stopped using such domes because their head brazier went to work for the Great Northern!)

Chapter 5

Fluxes

We have alread seen that the bond between parent and filler metal is metallurgical, and hence needs a chemically – or perhaps better stated, metallurgically – clean surface on which to form. All soldering and brazing is carried out well above room temperature, and most base metals will form oxide or sulphide coatings when heated – many will do so almost instantaneously even at room temperature. Even 'clean' air contains sulphur dioxide in urban areas and if torch heating is used the products of combustion may also react with constituents of both base and filler metals. If we could get a really clean surface and *keep* it clean (e.g. by making the joint in a vacuum) then no flux would be needed, but in the imperfect world of the normal workshop flux is, in the great majority of cases, essential.

Assuming that we had a perfectly clean surface to the base metal, and a perfectly clean stick of filler alloy it would be sufficient if we simply applied a substance which would exclude contaminants at the working temperature of the alloy. Common tallow works like this when plumbers make a soldered joint in a lead pipe. The lead is scraped clean and then coated with tallow. As lead is relatively slow to oxidise at room temperature this will keep it clean. When the heat is applied the tallow melts and continues to protect the surface. The molten solder is heavier than the molten tallow so that the latter forms a film over this too; neither the air nor the waste gases from the blowlamp can get at the solder below the flux, which remains until the joint has set, after which it can be wiped off if need be. If all soldering and brazing were as easy as this our difficulties would be considerably reduced! But for many soldering and all brazing operations something a little more robust is needed. Indeed, even the tallow mentioned just now does have some reactive power on contamination at the soldering temperature, though it can be left on the joint afterwards with no ill effect.

The requirements for a good flux can be stated fairly easily.
(a) It must be capable of removing, and preventing the further formation of, any contaminants which might prevent the metallurgical bond.
(b) It must be sufficiently tenacious to

remain in place both before and during the heating cycle. Preferably it should be of a nature which will not 'run all over the place', either before or during heating.

(c) It must melt to a liquid of sufficient fluidity for the flux to be displaced by the filler alloy as the latter penetrates the joint gap.

(d) It must either (i) be easily removed after the joint has been made, or (ii) be sufficiently non-corrosive so that no harm will follow if it is left on the joint indefinitely.

(e) Requirement d(i) means that the flux should be such that it will not materially change in properties as a result of the heating. (Some fluxes, quite acceptable for short heating times, become almost impossible to remove if heated for a long period).

(f) It should be non-toxic when normally applied. In their very nature, fluxes are likely to be hazardous if misused, but they should, within reason, be non-toxic in use or, at worst, be such that simple precautions can mitigate their effects (e.g. ventilation or the wearing of simple protective gear).

(g) It should be simple to apply to the joint. For this reason some fluxes (e.g. resin) may have to be mixed with an inert vehicle, whilst others may be dissolved in water or spirit.

Looking at these requirements it is immediately apparent that there can be no 'universal' flux, whether for soft or hard soldering. Some base metals like stainless steel have extremely resistant films, likely to be a nuisance in both soft soldering and brazing. Others form very easily removed films; in soldering tin-plate, for example, the flux may be

needed more to protect the filler solder rather than the workpiece. Similar considerations apply right through the list. One point may need emphasis, however, when considering requirement (c). It might be thought that a liquid flux would meet this in every case. This is not so, for the 'liquid' may well be no more than the solvent carrying the active medium — indeed, this is normally the case. The well known 'Killed Spirits' flux used for soft soldering for hundreds (if not thousands) of years is liquid, but it is the melting point of the *zinc chloride* — its main constituent — which matters. The additives usually applied to the home-made version (ammonium chloride, common salt, or tin chloride) serve the purpose of lowering the melting point from about 270°C (too high for most soft solders) to around 175°C.

Requirement (d) is of paramount importance in the case of electrical work and absolutely vital when dealing with printed circuit boards for electronic equipment. So much so that the materials of which the conductors on the PCB and the terminations of the components are made are usually designed so that a truly non-corrosive, protective flux can be employed. So far as the final point is concerned 'ease of application' can be as much a function of how you prepare the surface of the base metal as of the flux. A 'rub over with fine emery' may deposit so much oil on the surface as to prevent a water-based flux from wetting at all — even finger-marks will interfere with some liquid flux when soft soldering.

This leads me to a point which is, I think, well worth emphasising. In brazing especially it is true that an active flux will cope with mildly oxidised copper or brass (not so much with steel) and even at soft soldering temperatures a really active flux can do so as well. But it is hardly sensible to ask the flux first to clean the metals (filler included) as well as cope with the rapid oxidation which will occur during the actual heating cycle. A little time spent cleaning the parts beforehand is well worth while – in fact, on copper and its alloys a mild pickle beforehand will work wonders. But the actual cleaning itself must be done with some thought. The oily piece of 'service emery' at the back of the bench may serve well enough for coarse brazing, but even here other means are desirable. I prefer to use silicon carbide paper used water-wet (the so-called 'Wet-or-dry') for two reasons; there is no deposit of oil (which may flash off with no harm done at brazing temperatures) but more important, far less risk of abrasive particles becoming embedded in the work. In the case of soft solders embedded particles like this can cause a phenomenon known as 'de-wetting', in which the solder refuses to spread over the surface, and remains in blobs and globules no matter how much flux is applied. A far better material for soft solder work is powdered pumice and water, the material used by dentists to clean up your teeth!

FLUXES FOR SOFT SOLDER

(1) **'Inert' fluxes** On the face of it a 'non-corrosive' flux is a contradiction in terms, for if it has no chemical action it can hardly perform its office. However, there are substances which show mild reactions at the soldering temperature but none at room temperature and others which change their nature to some extent on heating. In general, however, it is best to regard all these fluxes as 'protective' – i.e. a clean solderable surface is necessary beforehand.

These fluxes are usually incorporated in the solder itself as a 'core' (or, better, as a number of cores) and any which are offered to conform to BS 441 can be relied upon to be truly non-corrosive and may be left on the joint indefinitely. Indeed, some (if not all) accord a definite protection to the joint, forming as they do a corrosion-resistant coating. Similar grades of flux can be obtained in paste form. These will almost certainly be labelled as conforming to a Ministry of Defence specification – DTD 599A – as well as to BS 5625, Grade 1. All such can be relied upon and may safely be used for electrical work, or in other situations where flux removal may be difficult. It is, of course, important that such fluxes be used only on surfaces which are either chemically clean or which have been pre-tinned.

The model engineer may well find that he has plenty of 'ordinary' flux but none in this class when a job turns up which needs it. All is not lost! I have already referred to the use of TALLOW, which older readers may still employ as a tailstock centre lubricant. This can be used as a flux for ordinary soldering, as well as for 'wiped joints'. It works very well on tinplate and equally so on lead-coated steel ('ternplate') if the latter is clean. It can be used on *really* clean copper and brass, but some rubbing with the soldering iron may be necessary as well, after which the tinned surface will take solder under the tallow with no difficulty. It is a low melting-point substance and is easy to apply.

Ordinary RESIN is an excellent inert flux and forms the basis of most of the commercial varieties. The main difficulty is that of application. When soldering large cables into termination sockets it can simply be powdered over the joint, but this is not always possible for fabrication work. Powdered and mixed with tallow it forms a simple paste type flux or it can be dissolved in alcohol (methylated spirit) and applied by brush. In this form it is very handy, as surfaces can be cleaned, coated with flux and then left for some time before soldering. I use resin almost exclusively for all electrical work and keep a lump handy to touch with the soldering iron when in service. When tinning wire or strip it is easy to heat this with the soldering iron whilst drawing the work over a lump of resin, and good tinning will result. Resin is more active than tallow at the soldering temperature but provided it is not subjected to excessive temperatures (well above soldering temperature) will remain inert and protective at room temperature.

(2) **Active Fluxes** These are, almost without exception, based on zinc chloride, with a proportion of other chlorides added to adjust the melting point. You may think this odd, in that they are usually a liquid, but of course the water forms no part of the fluxing action; this boils off and it is the solid salt remaining which must melt and act as a flux. Such fluxes will tackle almost any base metal and are effective even when the surface condition is not ideal. However, there may be a little difficulty with some grades of stainless steel, and for this class of base metal fluxes based on phosphoric acid are better.

Everything has its price, of course, and these fluxes will cause corrosion if left on the workpiece after soldering; they *must* be removed, and they should NEVER be used on any electrical connections. Nor should they be used if it can be avoided on any closed vessel where access to the interior may be difficult. In most cases it is, on the whole, easier to clean the surfaces so that one of the 'semi-active' fluxes dealt with below can be used rather than to have the chore of thorough cleaning afterwards. The problem is aggravated by the fact that these fluxes are waterborne, for as the water boils off it splatters the active flux over adjacent areas — (see Fig. 24). This can be a real headache in another direction, too. Such splatters can get on to the surroundings of the soldering bench and cause cor-

Fig. 24 *Corrosive effects of fluxes. Left, brass soft soldered with paste flux. Right, identical joint soldered using a waterborne flux. Both were wiped clean and left for 7 days before photographing.*

rosion of metal not involved in the actual soldering job at all. Naturally you must take precautions to avoid getting the fluid into your eyes or any cuts as well.

Such fluxes are usually purchased ready made – the well-known 'Baker's Fluid' is one – and there are others graded for particular metals, stainless steel especially, formulated by the solder manufacturers. Others are made to suit both the high melting-point solders and those of especially low melting point used for work on pewter and the like. These will be made to conform to BS 5625. Active fluxes can also be had – rather more convenient in my view – in paste form, the vehicle in this case being tallow or Vaseline. However, there are occasions when some particularly difficult joint crops up (especially in the repair of ancient machinery or household utensils!) and no such flux is handy. In that case an improved version of the old-fashioned 'killed spirits' can be made up quite easily.

The base is a 10% dilution of hydrochloric acid in water ('spirits of salts') into which zinc metal ('Granulated Zinc' at the chemists) is dissolved. This will effervesce, releasing hydrogen, so keep naked lights out of the way and treat the acid with reasonable care. Though dilute the acid IS an acid. Keep adding zinc until the effervescence ceases and there is some metal left which is not attacked, and then decant the liquid. This will serve as it is, but an improvement will be found if you add about 10 grams per litre of ammonium chloride. If you don't want to muck about with so much acid you can make up a proper flux which conforms to Ministry of Defence specifications, but you will have to buy the zinc chloride etc from the chemist. The mixture is: clean water, 1 litre, zinc chloride, 100 grams, ammonium chloride, 10 grams. When these are dissolved add (very carefully) hydrochloric acid at the rate of about 10cc (sorry – millilitres!) per litre. Frankly the cost of commercial flux is so little that it is hardly worth while, but I have found myself having to make some occasionally over the years.

(3) **Intermediate Fluxes** These are 'active' in the sense that they will react with mild oxide films, but are sufficiently non-corrosive to allow them to be left in place on many metals – it is only necessary to wipe off the surplus flux while the joint is still hot. The basis of such fluxes varies widely, but most are organic compounds and quite beyond the capacity of the amateur to make at home. They will be manufactured to BS 5625, Grade 3 and if in doubt the solder or flux manufacturer should be consulted; no trouble should be anticipated when these are used on any metal other than stainless steel and aluminium, but I would not use any on electronic work and only when unavoidable on ordinary electrical connections. Resin based fluxes are best here. They are available both in paste form and as liquids.

(4) **Aluminium Soldering Fluxes** The non-corrosive properties of aluminium derive entirely from the oxide film which forms on the surface almost instantaneously. It follows, therefore, that ANY aluminium flux is likely to be deleterious if left on the joint afterwards. The majority of those available are so formulated that they cease their activity, or the activity is much reduced, at temperatures well below the soldering temperature, but even so it is imprudent to allow them to remain. Fortunately they are easily removed with

warm water. It is, of course, possible to solder aluminium without flux using the 'abrasion method' mentioned later.

The 'Intermediate' grade fluxes are the backbone of all casual soldering and the sort usually stocked in local toolstores. 'Active' fluxes are needed on higher carbon steels and cast iron and sometimes on copper (especially if you are caulking already brazed joints) and special types for stainless steel, and all these *must* be removed afterwards. For tinplate the resin based fluxes are quite adequate, and nothing else must be used for all electronic and most electrical work. Over the years you will gain experience, but you can do a lot worse than follow the practice of using the mildest flux which will serve its office. I find that my tin of 'Baker's Fluid' is very seldom used these days, and plain lumps of resin seem to be used more and more! Above all, preliminary cleaning of the joint surfaces will do as much for 'good jointing' as will any exotic flux.

FLUXES FOR BRAZING

It is clear that at the high temperatures used in brazing we need fluxes which are not only of higher melting point than for soft solder; they must also be active at that temperature and will have to cope with considerably more oxidation both of the workpiece and the brazing alloy. Fluidity is important, too, for with gaps as small as 0.002 inch a viscous flux may not penetrate. There is another point, which cannot be emphasised too much, and to which I shall return later. As the alloy penetrates the joint it must push the flux in front of it, so that none of the flux remains inside the joint. A viscous flux may offer so

much resistance that the surface tension of the molten alloy is insufficient to 'drive' it. On top of all this the flux must be such that it can be removed afterwards without too much difficulty; in some cases this is the most serious problem of all.

The flux most commonly used before the introduction of the lower temperature silver brazing alloys was BORAX. This still forms the basis of modern fluxes for use on the straight copper-zinc alloys where the brazing temperatures approach 1000°C, and can be used by itself. The unpopularity of this material as flux amongst older users is, perhaps, due to a failure to appreciate its nature. If bought over the counter at the chemist it is in crystalline form, finely powdered. This means that the chemical compound is associated with 'water of crystallisation' – it contains water locked up in the solid crystals. This water is discharged when the material is heated to quite a low temperature and at this stage the grains swell to an enormous degree (Fig. 25); a pound of borax could swell to fill a bucket. If this happens in the joint the flux is almost always displaced from the joint and if grain spelter (or even brass turnings) is being used that, also, will be displaced.

There are two alternatives. The first is to use 'Anhydrous Borax' – identical material but not in crystalline form. The second is to 'prepare' the borax, and that is what I do. It is only necessary to heat the flux gently and allow it to froth up in some suitable container, stirring it about occasionally. When it ceases to be sticky and powders at a touch it can then be ground up again – after allowing it to cool, of course.

(Some practitioners take this process further, and actually melt the flux, again

Fig. 25 *Borax. Left, the powder, as bought. Centre, the same quantity after heating to about 120°C. Right, the same, heated to melting point. The 'frothing up' in the centre can displace both flux and alloy from the joint.*

taking care not to overheat it. The fused material is cooled and then ground with a mortar and pestle; this is supposed to give a better result, but it is not easy to get a fine, even powder.) The flux can then be stored ready for use. Subsequent mixing with water for application to the joint will not result in any serious frothing though the water will, of course, boil away as heating commences. I must confess that I seldom use borax these days, as I find that the high-temperature flux which I use for 'Bronze-welding' (SIFBRONZE FLUX) is just as effective.

Fluxes for Silver-brazing Borax is not really suitable for the lower melting-point silver solders. You will understand that the flux cannot start to perform its office until it has melted, and even then it must reach a specific temperature before the action becomes fast enough. Borax does not become active until it reaches around 750°C; quite low enough for the straight copper-zinc brazing alloys, but not for the silver-brazing alloys. The fluxes provided by the alloy manufacturers for the lower melting-point alloys are based on various FLUORIDES, and have an activity range from about 550°C up to 800°C. At the higher end of this temperature range most of them begin to emit unpleasant fumes and the working range is usually considered to be from 550°C to 750°C. Prolonged heating results in more complex reactions between flux and oxide which produce a compound which is often difficult to remove, so that both excessive temperature and long heating cycles should be avoided when using these fluxes – in other words, in any normal silver brazing work with the normal alloys.

It follows that a different type of flux will be needed for the brazing rods of higher melting point – those with liquidus above 700-750°C. It is also necessary to use a different flux when joints may be fluxed during the assembly of the parts but which may not receive attention until other joints have been brazed, if these pre-fluxed joints are likely to reach a temperature high enough to melt the flux.

For these low silver or wide melting-range alloys a flux based on a mixture of fluorides and borates is used. Unfortunately these are not very active below, perhaps, 700°C so that their use with the high silver or higher cadmium alloys is not advisable. Again, when brazing stainless steel, the oxides of which are exceptionally inert, a flux with a high activity at relatively low temperatures yet which will remain active at the higher brazing temperature is needed.

This all adds up to the same thing — horses for courses. The alloy manufacturers recommend specific fluxes for each of their alloys, and these recommendations should be adhered to. This does not mean that you will need a shelf-full of tins and jars, for there is a good overlap between grades, but you cannot expect to do good brazing over a range of alloys with but the one flux. The truly 'universal' flux does not exist, even for the normal base metals, and special flux is needed both for stainless steel and for aluminium and its alloys. It is also especially important to select the correct grade of flux when heating is likely to be prolonged, for even if more flux is added as brazing proceeds there is a risk that the flux within the capillary gap may become exhausted.

Forms and Application of Brazing Fluxes Traditionally fluxes were marketed in the form of powder, for mixing with water to apply to the joint, and into which a pre-heated brazing rod could be dipped to provide it with a coat. The usual rubric was that the mixture should be 'of the consistency of cream' but I am afraid that the cream of the supermarket era bears no comparison with that which could be had in earlier times! I prefer to mix to the consistency of *toothpaste*, and for many years I used old toothpaste tubes recharged with flux as a means of application to the joint. More recently (following a spell in hospital) I have used disposable hypodermic syringes, without a needle, for the same purpose. It will be found that if a little Stergene or similar detergent is added to the mix the paste will adhere to the work more readily.

However, I have for many years bought my flux mixed in paste form in those grades for which this is available. It is *far* superior to any which you can mix yourself; I suspect that it is 'milled' rather than mixed, and even after some years in the pot it only needs a very occasional addition of a little water to bring it to a very easily used consistency. It is not quite as convenient for 'rod dipping'. If a heated brazing rod is dipped into the pot it *will* pick up enough flux to protect the rod, but not sufficient if you wish to add more flux to the joint. So, I do keep a very small quantity of the dry powder as well. Unfortunately this paste comes in rather large containers — 1kg is, I think, the smallest — but it does not go bad and can be kept 'in the stores' with a little transferred as working stock in a screw-top glass jar.

More recently a special penetrating flux has been set on the market, in which the flux is mixed with an organic fluid; Johnson Matthey's 'Mattiflux' is typical. This is a highly active flux with a working range of 550-800°C, very fluid at the working temperature, but it has a shorter working life at the operating temperature than the conventional flux. It has three advantages: it is available in small quantities (it is in fact sold in small toothpaste type tubes!) It will penetrate a lap joint, even if applied after the joint has been assembled, far more reliably than will a paste flux. Finally, it will remain 'pasty' for years.

Flux-coated rods are now available. These are, I think, more appropriate to industrial use on the one hand, and to the 'Do-it-yourself' market on the other, rather than to the model engineer. In industry such rods will be economic, as they will be turned over from the stores frequently, and their use ensures 'quality controlled' fluxing in manual brazing, while in the D.I.Y. field the manufacturer can ensure that the user does not forget to 'flux his rod'. My own experience with such rods is that the flux coat tends to chip off in storage, and I am a trifle apprehensive about their absorbing moisture in a workshop which protects me from an average of 85 inches of rain a year! Their performance is excellent – but I would not recommend them if your normal usage of brazing alloy is such that they stay on the shelf for long periods between usage.

Finally, 'Flux-alloy paste'. I have referred to this in the section on brazing alloys. The one point to mention at this juncture is that when considering the use of this form of brazing material you must make sure that both the alloy AND the flux are suitable for the work in hand.

Brazing Flux Removal All the fluoride type fluxes can be removed in warm water provided the work has not been either overheated or heated for too long a period. PICKLING OF THESE FLUXES IS NOT NORMALLY NECESSARY, though immersion in a 5% sulphuric acid bath does clean up the workpiece overall if of copper etc; naturally, the whole of the job which has been hot will tend to oxidise. But if the flux cannot be

removed in hot water – I use a nailbrush on it as well – then it has been over-heated or been kept hot too long. In such cases the use of a pickle made up from 10% solution of caustic soda is recommended, though the sulphuric acid pickle is fairly effective.

The high-temperature borate based fluxes are rather difficult, and for these the 10% sulphuric acid is the thing to use, but if they have been overheated at all then mechanical means may have to be resorted to. Fluxes for the brazing of stainless steel can be removed by immersion in 10% caustic soda solution, followed by brushing in warm water. In most cases, if the work is

Fig. 26 *The author's small (4 in. dia.) stainless steel quenching tank. The lid covers the pan before the work enters the acid.*

Fig. 27 *The consequences of a splash of Zinc Chloride flux on steel bar-stock.*

allowed to cool to 'black hot' in air and then plunged into cold water the thermal shock will remove the crystalline flux residue, but you must consider the effect of such a rapid cooling on the work-piece as well! You can combine this with a pickle, by quenching in the pickling bath, but it is not a pleasant thing with either the acid or the alkali. I show in Fig. 26 a pickling chamber I use for smaller pieces. Made up of stainless steel sheet, welded up with the MIG process by the local jobbing welding shop, the design is such that the lid covers the liquid before the work reaches the surface. It is 4 inches in diameter, and contains enough pickle to deal with up to about 6oz (200gm) weight; but it could be made larger if the dimensions are scaled up.

Conclusion The function of the flux is to remove the oxides and other contaminants which may form during the heating cycle. They are NOT cleaning agents, and despite what may be said in some quarters it is unwise to rely on the flux to remove either dirt or oxide on the base metal. The less the flux has to do in this direction the better it can cope with the onerous duty during the heating cycle.

A wide variety of fluxes is available, and each has its proper place. There is no truly universal flux either for soft solder or for brazing. Users should never hesitate to seek advice from manufacturers – most have data sheets which will give all the information required. If, in an emergency, it is necessary to use one sort of flux for a purpose other than that for which it is designed – think on! It may involve more rapid heating, or perhaps working hotter than normal, or you may have to face more difficult flux removal. The principles I have outlined should help a little. Finally, all fluxes are to greater or less degree inimical to the human economy; I deal in more detail with 'safety' later, but even a non-corrosive material like resin won't do a child's tummy any good (nor yours, for that matter). Treat all fluxes as you would any other chemical – take care, and keep them out of reach of children. And – watch out for spills, especially when using the more active liquid soft-solder fluxes. Fig. 27 shows the catastrophic consequences of an un-noticed splash on some of my steel bar-stock!

Chapter 6

Soft Soldering Techniques

I do not propose to deal with the 'wiped joint' techniques used by plumbers on lead pipe and by cable jointers; these are covered at length in specialist books and are seldom needed these days. Nor is the dip or wave soldering process, used for printed circuit boards, of much concern, though I shall say a little about the manual soldering of components to such boards later – an increasing number of model engineers use radio control and if they make their equipment rather than buy it then a printed circuit will almost certainly be involved. This leaves the three techniques – soldering iron, sweating, and blow-pipe or torch work, and we will look at them in turn.

(1) **Soldering Iron** Also known as a soldering butt or soldering bolt in some parts of the country. Whilst iron CAN be used to solder with (I show a somewhat unorthodox tool in Fig. 38, page 64) the business end is universally made from copper. The copper has a high specific heat, and so holds a greater heat content than most metals, allied with good conductivity so that heat is rapidly carried from the body to the tip, where the work is done. We shall deal with 'selec-

tion' in Chapter VII, so for the present we will assume that an adequate method of heating the bit is used. The best compromise of working temperature is about 260°C, though with electrically heated bits this may rise to 350°C or more during periods of idle time. Naturally if a high-melting-point alloy is used higher temperatures will be needed; the important point is that the point of the bit must remain above the liquidus of the solder at all times. Naturally, no-one actually measures this whilst working (though thermostatically controlled irons are available for delicate work) and the usual test is to hold the bit an inch or so away from the cheek. If you do this at a time when the iron is working well you will soon be able to judge when it is 'right'.

The first operation is to 'tin' the working surface of the bit. The surface should first be filed smooth, removing any 'muck' or oxide if it is a new one. Don't use emery, as this will almost certainly leave particles of abrasive embedded, which will effectively prevent proper tinning. Rub the clean surfaces with flux – on a newly filed copper surface there will be no need for fierce flux and I use resin. Whilst the bit is

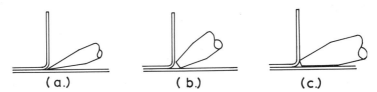

Fig. 28 *(a) A sharp-pointed bit can prevent solder from entering the joint. (b) Solder will leave the bit and penetrate the capillary gap on the workpiece. (c) The capillary between bit and work will 'hold' the solder and none will enter the workpiece.*

heating cut off a short piece of solder and set this in a tin lid with a little paste flux. As soon as the bit is hot enough to 'run' the solder apply it to the piece in the tin, melt it, and rub the faces of the iron about, one after the other. This will – or should – give a smooth, even coating of solder. You don't need to coat the whole of the copper – just the parts which will be used. You can then wipe the end with a piece of felt or 'moleskin' over which powdered resin has been sprinkled. The more care taken over the tinning of the bit the better will be the subsequent soldering performance, and even though a new iron may be supplied 'ready tinned' the process should be gone through – though it is not necessary to file it up in that case.

To make a joint, flux the faces of the base metal and heat the iron to working temperature. Flux the end of the stick of solder – don't dollop it on; just a thin film. Wipe the bit with the fluxy felt and then apply it to the stick of solder. This will melt and a blob adhere to the soldering bit. Set the point of the bit to the joint and slowly draw it along; the solder will pass from the bit into the capillary of the joint until there is none left on the bit. Then repeat. It is as easy as that! You can, if you wish, run the clean fluxed iron along the length of the joint a second time to even out irregularities, but if the iron has been hot and the joint clean and fluxed this ought not

to be necessary.

As I say, 'easy'; but it doesn't always work! One mistake made is in the shape and attitude of the point of the bit. Shape first (see Fig. 28). At (a) the bit has been filed to a sharp point, in the mistaken impression that this will give a sharper fillet. Instead, the sharp point penetrates the joint gap and effectively blocks it, so that the solder cannot flow past it. At (b), however, the solder blob, collected at the relatively blunt point, is immediately attracted into the gap by the capillary action. Naturally you must use reason – too wide a flat on the tip will leave the solder it carries out of contact with the work. But a small flat there should be. Now look at 28(c). Here the bit has been laid with one of the flat sides in contact with the work – on the assumption that this will heat the joint more rapidly. So it does, but it also provides a very hot capillary gap, which is bound to be hotter than that of the joint. So, the solder runs between the bit and the work instead of into the joint gap proper. There are, of course, occasions when we WANT to coat a surface, when the attitude of 28(c) is correct, but not when making a capillary joint.

You may find that the solder freezes in the angle of the joint before it can run into the gap. The bit has insufficient heat capacity to raise the work to the operating temperature. This may be due to a 'cold iron' but far more often the cause is insufficient mass in the bit. I shall be saying something about this, with particular reference to electric soldering irons, later, but it IS important to use a bit which is well up to the work. I always use the largest bit which can be applied to the work when making soldered joints (other than electronic) and even for delicate jobs it is better to use a heavy bit if you can. The alternative, that of getting the soldering iron hotter, is inadvisable – though you must, of course, have it around 250-275°C at least. Too hot a bit will 'burn' the solder, forming a heavy oxide film which the flux cannot deal with. This is less likely with electrically heated bits, but even there it can happen. The 'pulse control' on the 'Superscope' (see p. 65) is useful in this respect.

Remember, you have to melt the solder first, and a 65-watt soldering iron is not capable of melting more than 1 gram per second. Then you must heat up the work to a figure which will ensure that the solder will flow right through the joint gap before any of it freezes, and the greater part of the heat needed for this is transferred to the work through the solder itself. This heat comes from that stored in the bit. You cannot make a good soldered joint unless the bit, the solder *and* the joint gap are above the liquidus of the alloy.

There is an alternative method of applying the solder, using solder WIRE. This way the hot bit is applied to the joint and the solder wire then held against the gap and melted by the bit, after which it flows into the joint. It is an effective method but has certain drawbacks. First, it needs two hands, and there are many cases where a 'third hand' may be needed to hold part of the work in place. Second, you are melting solder continuously, and this causes a

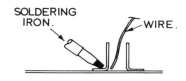

Fig. 29 *A case where solder is best applied in wire form, with the bit serving merely as a heat source.*

continuous drain of heat from the tip of the bit. Third, the use of flux-cored solder is almost essential, and unless you have wire of the right grade of alloy AND the right grade of flux a poor joint may result. The technique is ideal for wiring up electronic assemblies, and for 'spot joints' on ready-tinned parts of a fabrication, but I never use it for seams or capillary joints of any size. All too often the bit cools down in the middle of the job and you are left with solder, bit, and work all in one!

The wire technique can be useful in some special cases, of which a typical one is illustrated in Fig. 29. Here we have a joint into which it is not possible to get the point of the bit, yet it is not desirable to have a visible seam on the outer flange of the upright. The work is heated from the outside and the solder wire applied opposite; as soon as the work is up to temperature the solder will 'follow the heat' and flow right across the flange. I use this method on wide seams, when applying the bit in the usual way may not get the full width of the joint up to temperature, so that the solder sets before it has traversed the width of the joint. In this sort of situation I use plain solder wire, fluxed with one of the paste fluxes, rather than the resin cored type.

Tacking and pre-heating All too often we find we have two parts which will not 'stay put' whilst being soldered, and

Fig. 30 *Top, a long Tee-joint, 'tacked'. Below, the joint partially completed, between tacking points.*

which are impossible to clamp. In such cases the trick is to 'tack' the joint first, as shown in Fig. 30. The work can be hand-held while a number of small blobs of solder are set along the length. You don't need a lot of them – they should be far enough apart so that when the seam is run it will have set at blob 'a' before you reach blob 'b' and so on. You will find that the tacks will hold the two pieces together when running the seam, for the solder sets well before the gap between tacks is filled.

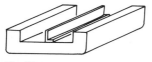

The heavy base of a joint like this will need preheat when using a soldering iron.

Fig. 31

Preheating is needed when soldering thin to thick pieces, Fig. 31. Even with the largest available bit it is difficult to bring such a thick piece up to the liquidus temperature and even if you succeed in doing so locally the solder will freeze so quickly that the joint will be very unsightly. On the other hand, if the work is done by torch soldering, as described later, the parts will have to be kept clamped for quite a long time because it freezes too slowly. The procedure is first to clean and flux the parts and prepare the bit. The holding arrangements must be worked out, too. The thicker piece is then gently heated with a torch or lamp (rest it on a non-conducting surface, of course) until the flux shows signs of becoming active. Or, if you like, until thin gauge solder wire just WON'T melt. The second piece can then be set up and you will find that you can make the seam in the conventional way with a reasonable size soldering bit; there is sufficient heat in the heavier part to call for only a small supply from the bit. Despite that, the solder will still set fairly quickly.

Pre-tinning With some materials it is very difficult to make a good joint directly. The solder is disinclined to wet the surface and, as we have seen, without this intimate contact between filler and base metal no proper bond can be made. The trick here — useful especially with steel and cast iron — is to use the 'friction soldering' technique which I shall discuss further when we come to aluminium. The joint surface is cleaned: if it has been machined it should be degreased with both chemical and detergent treatment. An active flux is applied and a 'puddle' of solder melted on to the surface. The well-tinned soldering iron is then applied and scrubbed on the workpiece under the solder surface. This is a case where the flat of the bit *should* bear on the work, as in Fig. 28(c). It will be found that after the surplus solder has been removed (it can usually be wiped or shaken off) the surface has 'taken the tin' quite satisfactorily. It pays to pre-tin almost ALL joints other than those on tinplate, german silver, and lead-coated steel. Fortunately the copper-based alloys are nowhere near as difficult as the ferrous metals, and it is usually sufficient simply to apply the flat of the bit, carrying a load of solder, to effect a clean coat.

Solder paint can, of course, be used for tinning — this is one of its main purposes. In this case there is no need to apply flux to the surface as this is contained in the paint. It is only necessary to spread an even (not too thin) layer and heat gently from below with a spirit lamp or similar. If direct heating is unavoidable then the flame must be gentle, and not be allowed to concentrate on one spot, but wherever possible solder paint should be heated indirectly. If you need to restrict the flow of solder to a definite area then a 'resist' can be applied. Black-lead — the stuff once used to polish up the old cast-iron kitchen ranges — should be painted on as a stripe around the area to be tinned. The solder will not cross over and the black can be removed afterwards quite easily. (It can be obtained made up specifically for the purpose, when it is known as 'Plumber's black'.) The tinned surface can be wiped clean of both flux and surplus solder with a rag, though I usually powder a little resin on the rag first. This will form a protective inert flux coating and make subsequent soldering operations much easier.

(2) **Sweating** A sweated joint is one in which both joint faces are tinned and then held together while heat is applied so that the two tinned faces fuse together. It is probably the strongest form of soldered joint, even more so if the two faces have been machined, and is very useful indeed when fabricating parts in which the joint faces are too wide to be sure that capillary attraction will carry the solder right across. Fig. 32 is an example.

The cover-plate is made from three parts, all of brass, and six brass washers forming the pads for the holding-down nuts. The central pad was filed to shape and after tinning both pad and cover and one side of the washers, the stud-holes were drilled. The washers are located by small sticks

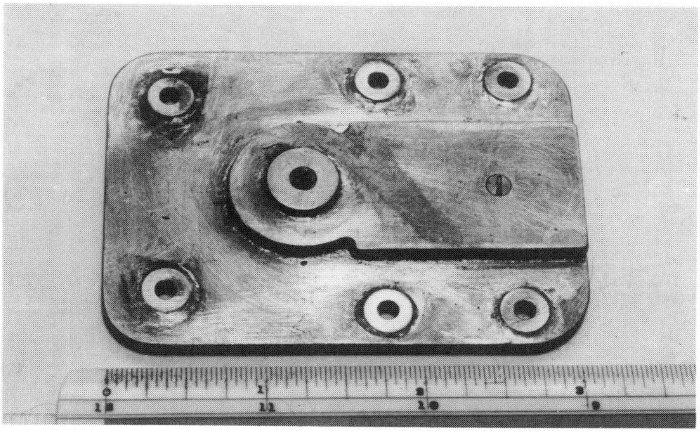

Fig. 32 *An example showing the use of 'sweated' joints.*

of wood, and the central pad by means of an 8BA countersunk screw which will not be seen when the model is assembled. The centre boss was, after machining, also tinned and held in place using (on this occasion) a very rusty bolt, to which solder would not adhere. The whole was then heated from below until the solder 'ran' at all joints. As the design called for a fillet around the centre boss a little additional solder was fed in here, using solder wire.

Reasonable pressure is needed for such joints. In the case just described this was provided by the screws, but parts may be clamped together or even riveted; as soon as the solder melts the faces will come together to form a very small joint gap. Another useful device is the spring clip used to hold bundles of papers. These will stand up to soldering temperature without losing their temper. The one point at which care is needed is to avoid sideways movement of the two parts – when fluid the solder is an excellent lubricant. A couple of small pegs will usually look after this.

Sweated joints in sheet material can be effected using the soldering iron as a source of heat. The largest bit available should be used and the work rested on a non-heat-conducting surface. Better heat transfer will be achieved if you see that there is a small quantity of solder on the bit.

(3) **Soldering Electronic Equipment**
This is a large subject and I can do no more than skim over it. But electronic devices are appearing in both the domestic and the model engineering field, so that a few words will not be out of place.

We are faced with several differences compared with 'fabrication'. First, the parts are small, often very small indeed. Second, the components are, in the main, very vulnerable to heat. Third, the slightest trace of any corrosion will almost certainly cause failure of the device after a while. Finally, 'stray electric currents' can damage many of the devices used, and as no electric soldering iron operating at mains voltage is entirely free from leakage this may cause problems.

I will deal with the last point when we come to look at 'the tools for the job'.

Corrosion is avoided by using nothing but resin and the special 'electronics grade' cored solder for all joints. Active flux should never be needed, as the component terminations are almost universally treated to make them easily wetted by solder. They often LOOK dirty, but a check with a clean soldering iron will soon prove whether or not the protection has been retained during storage. The one area in which difficulty may be found is with the solder tags used to conect components to screwed terminals. These are often simply pressed out of tinplated brass and I always re-tin these before use.

Departing from the actual soldering process for a moment, the one thing that constructors should always bear in mind is that there are few joints which you can be sure will NEVER need 'unsoldering'. It is a golden rule that where possible all wires and terminations should 'stay put by themselves' before soldering, but to crimp over a wire end to a perforated tag is asking for trouble later (see Fig. 33). The joint at 'a' will be very difficult to dismantle, but that at 'b' is almost as secure yet easily undone. Similarly with printed circuit boards ('PCB'). The procedure here is to assemble all the components of like nature to the board at once, when all these can be soldered before proceeding to the next batch. The components must be retained so that they do not fall out or drift over sideways. Fig. 34 shows how this should be done – the wire ends should *not* be bent over as at Fig. 33. The slight angle of the wires will hold the part quite securely, yet if the time ever comes for the component to be removed it can easily be pulled through the board after melting the solder. (In which connection, the avid enthusiast will, no doubt, have

Fig. 33 *(a) This wire connection will be difficult to detach even when unsoldered when component replacement is needed. (b) The preferred method of anchoring wire to a termination.*

invested in a 'DEsoldering tool' – simply a small soldering iron with a device which sucks the solder from the joint.)

So, to the actual soldering. A small, but hot, soldering bit must be used, so that the joint can be made very quickly; any prolonged application of heat may ruin the component. If using connecting wire, say to a mains transformer, tin it first, and if it is stranded twist the strands together first. Set these through the holes of the terminations as at Fig. 33(b), apply the soldering bit and then touch this with the solder wire. The solder should melt almost instantaneously and flow round both wires and tag. If it is reluctant, then either the bit is not hot enough or you are using too large a gauge of wire. For this sort of application No. 18 gauge is the largest (for cored solder) that I would use. There is no need to remove the flux if proper cored solder has been used.

For printed circuit boards an even finer bit may be necessary, especially if

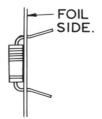

FOIL SIDE.

Fig. 34 *The preferred method of anchoring components to a printed circuit board prior to soldering.*

it is at all crowded. Having inspected the terminations of the batch of components to be fitted, and re-tinned any which may look suspicious (many resistors may look very 'dull' but are, in fact, intended to be that way) set these in place, spreading their legs as in Fig. 34. Start at one end and proceed exactly as for the previous example; bit applied to the wire and the copper foil, and the solder applied to the bit very close to the wire. In this case I recommend nothing larger than No. 22 gauge flux-cored solder, and preferably multi-cored stuff at that. Work quickly, keeping the length of time the bit is in contact as short as possible. On a board of any size you may have to give the iron a rest every so often, to enable it to recover. Then, having inspected all for poor or missed joints, cut off the sur-

Fig. 35 *Top, close-up of a soldered wire on a printed circuit board. Both wire and foil are properly wetted. Below, three 'Heat Shunts' applied to the terminations of a transistor.*

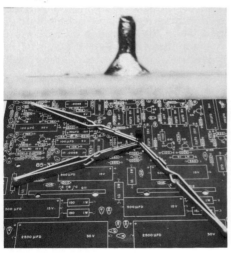

plus wires from the terminations, set the next batch of components in place, and repeat. By adopting this procedure you avoid finding such a mass of protruding wires that you cannot reach a joint, you check the soldering as you go on, and you give the soldering iron periodic rests to recover heat. The soldered termination should look like Fig. 35A, with a smooth curve from wire to P.C.B. foil.

There are some components – transistors, diodes and integrated circuit blocks, as well as some types of capacitors – which are VERY sensitive to heat. For such components I recommend the use of a HEAT SHUNT. This can be no more than the nose of a pair of snipe-nose pliers, gripping the wire on the component side of the PCB whilst the other end is soldered. This is effective enough, but you will find you need three hands! I use proper heat-sinks, as shown in Fig. 35B. These clip on to the wire termination and stay put, and you will have little difficulty in getting four onto one transistor. They are quite cheap, only a few pence each, or can be made from spring brass.

There is nothing difficult about this class of work. A little practice on dud components is well worth while if you have done none before. The most common faults are DRY JOINTS, due to imperfect wetting by the solder. This is the most common cause of 'noise' in a new piece of circuitry. COLD JOINTS, in which the solder is rough and knobbly, caused as a rule by over-working the soldering iron. If it doesn't melt the solder wire almost instantaneously it is not hot enough and should be given a rest. Another cause is a dirty bit, which reduces the heat flow from bit to work enormously. Too large a gauge of wire may be responsible, but the main

Fig. 36 *A fairly simple example of printed circuit work. Note the clean surfaces of the soldered joints after 20 years of work. Resin cored solder was used.*

reason for using a smaller gauge is to keep the amount of solder in proportion to the joint. Another fault is a BURNT JOINT, where the bit has been applied for too long, resulting in the copper foil parting from the substrate of the PCB. Finally, SOLDER BRIDGES, where too much solder having been applied it has overflowed to form a conducting path between adjacent strips of foil. To remove these (in the absence of a desoldering tool) flux a piece of *clean* stranded wire (that from a fairly heavy piece of mains flex will do) heat the offending solder and apply the end of the flex. Most of the surplus solder will be drawn up into the bundle of fine wires. The process may have to be repeated. Fig. 36 shows one of the P.C.B. in my oscilloscope. Although it is just over 20 years since I made it the solder joints are still bright. Resin-cored solder was used.

(4) Torch or Blowlamp Soldering The mouth-blown gas blowpipe used to be an essential tool for all silversmiths, pewterers and the like. This was used to heat the workpiece, usually held in

clamps, and fluxed solder wire (of the appropriate type for the job) applied. It is rare indeed to find this skill used today, but the air-blown blowpipe is still an essential tool for all delicate work. Fig. 37 shows one such of 1000 BTU/Hr max. capacity (I deal with it in detail on page 68), and you will see that it stands on the bench, leaving the operator with both hands free to manipulate both

Fig. 37 *The Adapto needle-flame torch.*

work and solder. Flux-cored solder is used, of course. I find this tool of great help when soldering up very small components – valve handwheels etc. If fairly thick solder paint is used the manipulation is eased still more. For slightly larger jobs the normal gas/air blowpipe or a small blowlamp can be used, but it is seldom that the full-size 'brazing torch' type burner is needed for soft soldering. The part shown in Fig. 32 could have been made this way, by applying solder wire after bringing the work hot enough, but there was a little doubt about penetration into the gap. In general, apart from the pre-heating mentioned earlier I find the soldering iron more convenient than torches except when very small details are to be made.

When using a torch during soft soldering processes, even when doing no more than preheating one part before using the conventional soldering iron, there is considerable heat stored in the work. The alloy will cool down but slowly and will remain pasty for some time. It is important not to disturb the joint until the whole of the work is well below the solidus temperature.

(5) **Aluminium Soldering** Basically there is no difference between soldering aluminium and any other material. The correct grade of solder and flux must be used. The liquidus of the alloy is slightly higher than that of ordinary solders but still well within the capacity of a good bit. The thermal conductivity of aluminium is twice that of brass and five times that of steel (e.g. tinplate) but only two thirds that of copper, so that although a bit of reasonable size is desirable it need be no larger than one you would use on the same thickness of copper sheet. Modern fluxes make this work relatively easy but I confess that I still adhere to the older process of 'abrasion tinning' for joints that matter. The problem, as you will appreciate, is that aluminium forms a protective oxide very quickly, and the very coating that gives it its attractive properties also prevents the solder from wetting and bonding to the surface.

The procedure, already described but repeated to save you turning back, is first to clean the surfaces and a slight roughening (using wet silicon carbide, not emery, paper) does no harm. The surface is fluxed, as is the bit and the solder stick, and a puddle of solder melted on to the work. (This does *not* mean a solder *lake*, but sufficient to cover the work when all is tinned.) The hot bit is then rubbed on the workpiece surface under the solder coat and you will find that gradually the pool will spread to a proper film all over. The bit 'abrades' the oxide coating and the overlying solder pool prevents it from reforming. Once tinned (the solder itself will be a tin/zinc alloy) the joint can be made, using the solder stick and the same techniques as applied to ordinary solders. A few of the high-tensile aluminium alloys may present difficulties, but there are special grades of solder and flux for these cases. Frankly, you need have few fears over soldering aluminium for the ordinary type of work we do, but it does need just a little practice, as the alloy behaves rather differently from the tin/lead type.

(6) **Soldering Plated Parts** Bright plated NICKEL is almost impossible to solder. If unavoidable, then a very active flux must be used and the 'abrasive' technique just mentioned applied. A rub over the surface with silicon carbide paper may help, but though it is, with

difficulty, possible to repair such an article the strength of the joint will not be great. If strength is imperative then there is little option but to remove the plating, solder to the undercoat (usually copper) and then replate. CHROME PLATE cannot be soft soldered. SILVER PLATE is easily soldered, but the silver adjacent to the joint will tend to dissolve in the solder and may lose its sheen. The plating must be cleaned, for though silver does not form an oxide easily it does react with the sulphur in the air and this sulphide is solder-resistant. If clean, a non-corrosive flux can be used. CADMIUM PLATING can be soldered with a medium to strong active flux, but all too often such plated items (screws and the like) have been treated to reduce their surface activity. If this surface 'passivation' is first removed soldering is practicable, but the joint may look a bit rough. ZINC PLATED parts come in the same category as cadmium, but galvanised steel, if not too old, can be soldered reasonably well using an active (acid) flux. Galvanised sheets of any age must be cleaned abrasively down to bright metal first.

(7) **Other materials** STAINLESS STEEL will present few problems, especially if the 'solderable grade' (used for making chip-ranges etc) is employed. You are best advised to use the special flux made up by the solder manufacturers, but ordinary grades of solder will serve. You will have some difficulty with some of the stainless steel rod supplied for making engine parts and these are best regarded as 'non-solderable', but the main problem (apart from the need for specially active flux) is that the steel has a poor conductivity and the far end of the joint gap may not get hot enough to allow sufficient capillary flow to fill the joint. There are so many types of stainless steel on offer that it is impossible to generalise, but for tank-work the 'solderable' grade is perfectly satisfactory and is just as 'stainless' as any other.

CAST IRON is usually ranked as an 'unsolderable' material, but joints *can* be made if the work is pre-tinned using the abrasion technique. The difficulty is caused by the graphite (free carbon) in the metal, so that 'white' or chilled cast iron is easier to handle than the machineable grey. Any joints made on cast iron should be regarded as being rather poorly bonded, but I have made good joints by pre-tinning with pure tin, after sand-blasting the iron surface to remove surface graphite. A joint made with ordinary solder and Baker's fluid is quite adequate for securing chucking-pieces to small castings. Of other ferrous materials, CARBON TOOL STEEL (Silver Steel etc) is easily soldered, but HIGH-SPEED STEEL is not. Indeed, it is almost impossible to persuade most grades of the latter to tin at all. GERMAN SILVER (Nickel Silver) is the solderer's dream, provided that it is clean, and for this reason I always used it for making the bodies of locomotives when I was building 3½mm scale models many years ago.

(8) **Flux removal** Flux of the resin type can usually be left in place, but if (to permit painting, for example) it has to be removed it is probably easiest to do this mechanically; the residue is fairly brittle. Warm turpentine is rather better as a solvent than the alcohols. The paste fluxes are usually soluble in hydrocarbon oils – paraffin and petrol – but washing with water and detergent afterwards is only prudent. In most

cases means of removal is stated 'on the tin'. The very active fluxes are the most difficult. For most purposes thorough washing in HOT water will serve, but this cannot be guaranteed to remove all the active elements. If complete freedom from subsequent corrosion must be guaranteed then the work should first be washed in *warm vinegar* (or a 5% solution of acetic acid) followed by vigorous brushing in hot water to which Stergene has been added. This may seem strange, as the flux was dissolved in water in the first instance, but once this solution has been heated the chemical composition changes. Merely to wash it will remove the *unused* liquid flux but may well make the secondary compound MORE difficult to get rid of. Treatment with the mild acid (5% citric acid, as used in wine-making, will also serve) leaves the secondary compounds water soluble.

The biggest problem arises when closed compartments result from the fabrication, access to which may be difficult afterwards. Even a small amount of aggressive flux trapped in a closed chamber can do untold damage. In such cases, and when access may be through very restricted apertures, the best plan is first to tin all the joint faces, using active flux if need be. Then to remove all traces of flux by the process just mentioned. And finally to complete the job as using the non-corrosive resin flux which can safely be allowed to remain inside the cavities.

Conclusion

Manual processes of soft soldering involve SKILLS, and practice is needed to acquire real proficiency. Modern solders and fluxes – and, as we shall see later, modern soldering tools – have made things much easier, so that even the relatively unskilled can make a tolerable joint. However, practice *is* needed for more complex fabrications and especially when these involve dissimilar materials or dissimilar masses of metal. No-one need be ashamed of practising on pieces of scrap (indeed, how is anyone to know?) and an hour or so occupied in this way will save many hours of frustration later.

Soft soldering should NOT be regarded as a 'poor relation' of welding or brazing. It is strong enough for many purposes – those who glue on their model locomotive wheels might care to reflect that soft solder is four or five times as strong. The soldering temperature is low, so that there is very little risk of affecting either the properties or the heat treatment of the workpiece – even carbon steel drills can safely be soldered into an extension rod. The cost is low and the equipment needed negligible; you can soft solder with a poker, or with the spirit lamp from a toy steam engine. And, a property not to be disdained, it is usually possible to dismantle and re-assemble soft soldered joints leaving no evidence that they have been disturbed at all. Finally, it is quick, especially for electrical connections. You can solder up all the leads to a mains transformer in far less time than it would take to undo and retighten a set of nuts and washers, AND you will be sure that they won't come undone!

Chapter 7

Tools for Soft Soldering

A good soldering iron will have a high heat capacity, so that sufficient heat is stored to complete the joint, and good thermal conductivity, so that heat is transferred rapidly from the body of the bit to the point; it will be relatively unaffected either by the solder or the flux, be comfortable to handle and be shaped to suit the type of joint being made and, finally, be of a type suitable to the work normally done. To clear this last point first, those who use soldered joints once every Preston Guild* for odd jobs would be content with a medium wattage electric tool, whereas one who did a great deal of tank-work would probably use a fairly large gas-heated pair of irons and someone whose soldering was confined to the making of radio control equipment would use a bit which would not even warm a kettle bottom if such needed repair. Horses for courses, and you must decide that one for yourself.

As I have already remarked, you *can* solder with anything that will be wetted by the alloy. In Fig. 38 you see an ordinary poker, tinned at the end, with a seam soldered up using it. Not a very good joint, rather rough. The 'bit' is the

*Local vernacular, equivalent to 'once in a blue moon'.

wrong shape, there is insufficient heat capacity, heat transfer to the point is poor, and the whole tool is awkward to use. But it did work after its fashion. However, the obvious material to use is COPPER and this is almost universal. It does not meet the third requirement exactly, though. Copper is dissolved by the tin in the solder at the operating temperature and the working surface of the bit will slowly erode away (see Fig. 7A, page 13). Periodic re-dressing is necessary, followed by re-tinning. Some manufacturers coat the surface – usually with iron – (virgin iron, not 'steel') but even this is not entirely effective. Industrial concerns in which manual soldering is going on 8 hours/day use a special type of solder which is so saturated with copper that it cannot absorb any more, but this is not open to the model engineer, though the cored solder known as 'SAVBIT' is reputed to be of this type. The problem is not serious – a few strokes of the file occasionally, that is all.

I have already remarked on the shape of the 'point' of the bit (see Fig. 28, page 52). The *weight* should be as large as possible for fabrication work, but you do have to hold the thing and too heavy

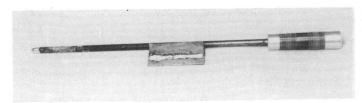

Fig. 38 *An ordinary light poker, cleaned and tinned, can serve as a soldering iron!*

a bit can be very tiring – you must effect a compromise. Again, the balance and shape of handle are very much a matter of choice, but is worth taking a little time when buying a new one. In which connection the mains cable of electric irons can be more than a nuisance. Some have a stiff PVC covering which has a mind of its own and keeps getting in the way. The flex should be what it says it is – flexible.

Fig. 39 shows 'ordinary' soldering irons of our sizes. At the top is my 1½-lb 'heavy' – just a little too heavy for most work (a 1-pounder would be handier) but ideal for large jobs and seams in thick material. Below that is a ½-lb bit, large enough for seams in tinplate but a bit light for copper. Below again a small 4-ounce bit, for light work. These three are all of the traditional type with the bit riveted to the stem, by far the best construction. All have rough, but com-fortable, handles. At the bottom is a small 'hatchet' bit, home-made from a piece of copper bus-bar. This shape is much the best for seaming, as the bit is drawn towards the operator instead of being traversed sideways.

In Fig. 40 we have three electrically heated bits by SOLON. At the top a large 125-watt, with a working bit weighing just over half a pound. At bottom, a similar one but 65-watt, 5 ounce bit, and in the centre another 65-watt with a detachable pencil bit, and a special point alongside, bent from copper rod, to get at an awkward spot. It is not easy to equate these in performance to those in Fig. 39, but we shall be discussing the heating of bits in a moment. Fig. 41 shows two very small ones with tiny bits for electronic work, 25 and 10 watts, both working at mains voltage. However, for all work involving 'solid state' devices, it is unwise to use

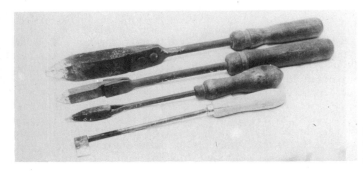

Fig. 39 *Classical soldering irons. Top to bottom, 1½ lb. bit, 8-oz bit, 4-oz bit, and hatchet shape.*

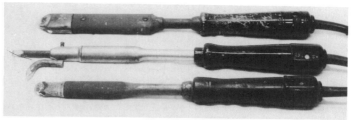

Fig. 40 *Three electric soldering irons by Solon. Top, 125 watt; centre, 65 watt with interchangeable bits; below, 65 watt of conventional type.*

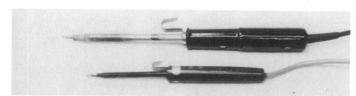

Fig. 41 *Two small electric soldering irons. Top, 25 watt, for instrument work. (Solon) Below, 10 watt, for electronic work. (Litesold)*

direct mains-voltage on the heaters and low-voltage units specially made for the purpose are desirable.

Fig. 42A shows an unusual one of this type – the 'Superscope' made by Scope Laboratories of Australia. The transformer supplies current at about 3½ volts, one lead connected to the body of the tool (and hence to the copper bit) the other to a carbon contact inside the body. When the lever is depressed (or, on the larger one, the ring pushed forwards by the thumb) the carbon and copper make contact and heat up is very rapid indeed – the working temperature is reached in 3 to 5 seconds. If the ring on the larger one is kept depressed the heat output is about 140 watts, and a temperature of almost 500°C can be maintained. By 'pulsing' the control you can reduce the output to anything you please – as low as 20 watt with a bit temperature of 200°C. The smaller of the two shown (the 'Miniscope') is rated at 10-70 watts, and has an even faster heat-up time and the same facility for controlling the bit output. Both bit and carbon have reason-able lives, and replacement is very easy.

This instrument was designed for electronic soldering originally but I have found that the larger one can

Fig. 42 *Top, Low-voltage Superspeed and Miniscope soldering tools with their transformer. (These have been in use for about 30 years) Below, the current design of 'Scope' soldering iron and transformer. (Courtesy Greenwood Electronic Components Ltd, Reading).*

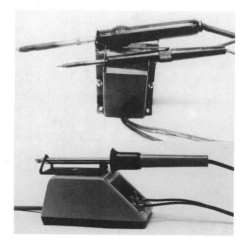

tackle quite hefty seaming work and light fabrications too. I bought mine about 30 years ago when I restarted playing around with radio transmitters, but very quickly transferred it to the main workshop. Fig 42B shows the present model, basically the same but rather more ergonomic. They are obtainable by direct order from the importers, AGB Southern Cross Industries Ltd, Cell Farm, Church Road, Old Windsor SL4 2PG.

Heating Methods I cannot be the only one to have started their 'soldering career' by heating the bit in the kitchen stove! This is quite legitimate provided the non-smoky part of the fire is used and that you keep an eye on things — it is all too easy to find the bit red-hot. If you are careful the bit will need no more than a wipe over with an old oven-cloth before use, but in the old days the adjacent syrup tin half-full of 'killed spirit' was used to clean and reflux each time. A more refined method was to set a short length of 1½ inch gas-pipe in the fire and set the bit inside this, to protect it from the sulphur in the fuel.

For ordinary soldering irons — Fig. 39 — the most effective heating method is to use gas. Serious practitioners will have two bits on the go, one heating while the other is in use, and Fig. 43 shows the little soldering stove I used when I was on town's gas. It must be over 50 years old now, but unfortunately the burner is of a type which could not be altered to work on bottled gas. It is lined with firebrick and when properly adjusted will hold two bits up to 1lb weight at about 300°C indefinitely, with no tainting of the tinning. Such stoves are still available, to suit modern gases, and the fuel consumption is very small indeed.

Heating with a brazing torch is most uneconomic, especially on propane, but hatchet bits can be heated very nicely on a paraffin blowlamp with a clip to hold it in place on the nozzle — indeed, at one time this clip was a standard fitting on most ½-pint lamps. In using such (or a brazing torch) the flame should be directed at the *body* of the bit, not the point. If you can rig up a firebrick 'cave' or a small pipe furnace like that shown in Fig. 44 (described in detail in 'Hardening, Tempering & Heat Treatment') a very small and gentle flame will suffice to keep the bit to temperature.

Electric soldering irons are carefully researched so that the weight of the copper in the bit is balanced to the size of the heating element. It must be remembered that the actual melting of the alloy and the forming of the capillary flow is done by drawing on the heat stored in the bit; the heating element simply backs this up, and reheats the bit between solder-strokes. The rate of heat release from the bit storage can be ten times or more the rate emitted by the element. When working on copper or brass the conductivity of the work is such that heat is carried away from the joint very quickly. It really is impossible to generalise, still less to 'work it out' when considering the wide variety of jobs which appear in a model engineer's or a jobbing repair shop. My experience suggests that the 65-watt tool is well enough for tinsmithery and 'Do-it-yourself' repair jobs, but for fabrication work and coppersmithing the 125-watt tool is almost essential.

The 'Scope' soldering tool works in a different fashion. The heat stored in the bit is very small and the whole of the energy needed comes from the electric current. However, this energy is applied, at very high intensity, only a few millimetres from the working point of the bit and the heat flow to the bit and from bit to work is very fast. In the short term it is about the equivalent of a 1-lb soldering iron, but it cannot keep that up for long and for continuous work must be 'pulsed' as previously described.

I have used both electric and gas-heated soldering irons all my model-making life, and fire-heated ones before

that, and there is no doubt at all in my mind that the twin-bit gas-heated arrangement is by far the best for any serious work. The electrically heated ones are, of course, cleaner, take up less room and can be plugged in anywhere. They take about the same time to heat up as those in the gas furnace. The 'Scope' is, of course, ideal for emergency jobs in this respect, as it comes to heat in seconds.

As to size, I suggest you consider nothing less than 125 watts for normal work, with perhaps the 65-watt with interchangeable bits for more delicate work. Those below this size are for instrument work only, and it is pointless to have a mains-driven tool when there may be transistors or integrated circuits about. Looking back, 65% of my work (on models) is done with the larger 'Scope' and the rest either with the 125-watt electric or one of the 'copper-bits' of Fig. 39. All electronic work involving 'solid state' is done using the 'Mini-scope', but the smaller of the two bits in Fig. 41 is used on much of my gear which still has valves, as it is very small in diameter and can get into difficult

Fig. 43 (opposite) A gas stove designed for heating two conventional soldering irons.

Fig. 44 This small home-made tube furnace is normally used for heat treatment, but with a small burner makes an excellent soldering iron heating stove.

Fig. 45 *The 'Soudogaz' blowlamp, with soldering bit attachment.*

Fig. 46 *The author's ancient spirit lamp.*

places. But I show in Fig. 45 a little chap which does have its place. This is the Soudogaz blowlamp with soldering bit adaptor. Provided you remember that the flame projects in front of the bit this is very useful for those outdoor jobs which crop up from time to time.

Soldering Lamps and Torches The classical mouth blowpipe is used with a spirit lamp, and I don't propose to go further into it than that. If you can use one you need no words from me, and if you can't (or haven't), no words of mine can show you how; you must learn the hard way – blow it and see! I will add only that in skilled hands it is a very precise tool indeed. However, the spirit lamp by itself has its uses for providing gentle heat when tinning, or for delicate work. The flame is very clean and deceptively hot – quite large pieces can be brought up to soldering temperature. My own, Fig. 46, is very old but similar ones are still sold by horological tool suppliers. Naturally, any spirit lamp will do (though I suggest a wick no more than ¼-inch diameter) and they are easy to make, but I like the glass ones as I can see how much spirit is left. The glass cap prevents too much loss by evaporation when not in use, but it is best to empty them when not likely to be used for some time.

Any of the smaller brazing torches can, of course, be used for applying general heat to a large component before soldering on a smaller part. Use a 'soft' flame. For pukka blowpipe soldering, however, you need something rather more delicate. Fig. 47 shows one such, by 'Flamefast'. It can be used as shown in Fig. 37 as a fixed flame with the burner set in one of two attitudes on the base (as shown, or dead vertical) or as a hand-held burner. It will run from any butane regulator supplying gas at 11 inches watergauge pressure (28 millibar) and needs but a small amount of low-pressure air. However, it is normally supplied with a tiny blower – actually one designed for supplying air to fish-bowls – and an adaptor to draw gas from the Taymar or similar disposable butane cans. It is a very handy burner indeed, as having control over both air and gas any type of flame can be produced, from a gentle 'warmer' to one of pinpoint intensity. It is just the thing for work which demands two hands to hold it.

'Third Hands' Many complex fabrications need jigs to hold the parts together whilst soldering, and I must confess that I find it a little odd that modellers who will quite happily spend hours, if not days, making up a jig for some machining operation disdain to do so for the equally important metal joining operations. Almost all industrial soldering, automatic or manual, is done using jigs and there is no reason why we should not do so as well. However, there are many cases where things have to be 'hand held' and you find you haven't enough paws to go round. I have in the past made up many gadgets to overcome this problem, but recently acquired the device shown in Fig. 48. This, which has a good base to sit on the bench, carries two clips which can face 'every which-way' and within the limitations of their grip will hold delicate work in almost any attitude. You see it holding copper pipes after making a 'U' junction to a tee-pipe. It can be had with a X2 magnifying glass, which helps with very small items – or when your eyes are getting on a bit. A device with a more positive grip, really for

Fig. 47 *Another view of the Adapto needleflame torch with support base and the 'fish-tank' blower at the back. (Courtesy Rhodes-Flamefast Ltd)*

Fig. 48 *A 'third hand', for light work, with magnifier. Care has to be taken as the clips are only spring-loaded, but it serves for soft soldering.*

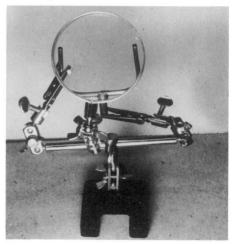

brazing, but equally well adapted to soldering, is shown at Fig. 68, page 97. A description of the manufacture of this one is given in 'Simple Workshop Devices'. It is a bit slower in setting up than Fig. 48, but the grip is much stronger.

Conclusion The soldering iron is a craftsman's tool and is, therefore, a very personal one. One man's perfect bit may seem quite unwieldy to another. So, in choosing one, keep in mind the purpose for which you need it; tend to go for the larger bit rather than the smaller (except for electronic work) but in the end, the one that 'feels right' is the one for you.

Chapter 8

Brazing Techniques

Before dealing with techniques proper I must reiterate some remarks which I have already made and, perhaps, emphasise them with a quotation from BS 1723 'BRAZING'. 'An essential part of satisfactory brazing is that the joints . . . are clean immediately prior to brazing. All surface scale, oxides, grease, oil and dirt shall be removed . . .' Note the 'shall' and remember that British Standards are MINIMUM requirements for approval. It is true that the flux will remove oxide films, but why give it more work to do than necessary? Further, flux combined with oxide forms *slag*, and that is but one form of dirt. Whatever people may say, preliminary cleaning is as much part of the brazing operation as is the deposition of the brazing alloy. I know that you can 'get away with it', and I have done so myself when circumstances were against me, but we are not in this business to get away with things — rather to make good joints, and the preliminary cleaning doesn't take much time. There is a further point — that of joint contamination *during* brazing. How often have you read 'Set the work in the hearth and apply heat . . . Now turn the job upside down and deal with the bottom joint, starting at . . .'? Now, that 'bottom joint' has been fluxed, and even if it hasn't melted when the top was heated it will be pasty, with water if nothing else. Did you clean the brazing spot before you started? Can you be sure that the flux hasn't picked up a piece of firebrick, asbestos, or whatever? Five minutes spent in cleaning a joint which may last 100 years or more is not too much to ask, surely?

The second point I want to make is the need for prior thought about handling the work. 'Pick it up with the tongs and turn it over'. Quite. But where are the tongs? Further, how are you going to grip the thing, remembering that it will be at or approaching red heat and will have no strength worth talking about? Oval boiler shells will probably work alright, but they look wrong somehow. With small workpieces you may find that applying the brazing rod just pushes them about all over the brazing table, too, so think about that as well. (I shall have something to say about joint design later.) However, it is not only tongs, but all the other likely needs; flux, an extra brazing rod in case you drop the one, something to relight the burner with in case it goes out, and so on. It is more than a nuisance if some little thing is found missing part way through the work!

Spelter Brazing In the old days this was the only form of brazing, using brass wire or spelter grains, usually the latter. The heat would be applied from a blacksmith's forge or, in a proper brazing shop, from a similar but specially shaped one. For small work charcoal fires would be used, with a very gentle draught. The dome-cover or pipe or whatever would be cleaned, and the spelter grains (which could be had in several degrees of fineness) mixed with flux which would be applied along the joint with a 'spoon'. The heat was then applied and in due course the flux would first swell up and then melt and finally, the spelter would melt also. This was recognised in the old days by the 'puff of white' which emerged as a tiny proportion of the zinc in the spelter burnt off as oxide. Now, the flux (borax as a rule) would often carry the spelter grains away from the joint or bring them together in lumps, so that it was usually necessary for the brazier to 'scrabble' the joint with a piece of pointed iron wire so that the discrete blobs of spelter 'found' the capillary gap and ran into it. Occasionally the frothing up of the flux would leave a small area unprotected and the spelter would not wet it: again, the pointed wire was brought into use to abrade the scale and give the flux a chance to work. This is the method I followed, even though heating with a torch, when using borax and brass turnings, but nowadays I find that modern fluxes work so well that the scrabbling is seldom needed.

When using brass wire, a modern high temperature flux, and torch heating the procedure is much easier. (You must remember that the old braziers were dealing with pretty large parts – the exhaust bend of a triple-expansion marine engine was quite a size!) The flux is applied to the work, not forgetting *between* the joint faces, and in reasonable quantities. Too much is far better than too little, but don't smother the thing. The end of the brass wire is heated a little and dipped into powdered flux, so that it is well covered for a few inches. This is both to protect the rod from oxidation *and* as a means of applying a little more flux during proceedings if this proves to be necessary. If the rod is short – 18 inches or so – I always flux both ends AND both ends of a 'spare rod' as well, but you may find it necessary to bend one end about a bit, to form a handle as it were, for brazing rods can be awkward things to manipulate.

Apply heat gently at first, to boil out the water from the flux paste, and try to do this with indirect heat – apply the flame to the body of the work, not the joint itself. If during this stage the flux retreats from anywhere apply a little to that point from your fluxed rod. As soon as the flux has settled down, increase the heating rate. If the work is of any size heat the whole until it gets pretty hot and then manoeuvre some insulating bricks around all but the joint area to retain the heat, or use an auxiliary burner (see page 79). Now transfer the torch to the joint area and, if you have that sort of burner, change from the 'broad blowlamp' type of flame to a hot concentrated one. Apply this to one end of the joint with the axis of the flame pointing along it. When it reaches brazing temperature (and the only way of knowing this is to 'try it and see' if you haven't done this sort of thing before;

for spelter-brazing it will be really red-hot) apply the end of the rod. This should melt and run into the joint. Keep the flame moving ahead of the rod and the rod just behind, but *as soon as* you have consumed all the fluxed part of the latter, *dip it in the dry flux powder* again. You MUST keep the end of the rod well fluxed. The melted alloy will always tend to flow to the hottest part in any capillary gap, so that the proper place to apply the cone of the flame is at the far side, not the mouth of the joint – Fig. 49A – though you may, of course, have to heat the mouth first to get the alloy to melt. If heat can be applied as at Fig. 49B, this is better still. Once the joint is run you may (when using brass spelter or rod) find it desirable to run over the joint again with the heat, but do make sure that it is still covered with active flux.

You may find that the rod doesn't want to melt, and all that happens is that the end of it runs into a little blob. This is a common error – you must heat the WORK, not the rod. But sometimes there is a definite resistance to melting even though the job seems to be at the proper temperature. A little reflection will reveal the cause. The rod has to be brought up to the 'liquidus' tempera-ture and then it needs the addition of the latent heat of fusion before it an actually melt. But the heat transfer from the hot body of the workpiece to the end of the wire is very poor indeed. Now, it may be that you are using too large a diameter of wire for the size of the job; I always like to err on the small side, even though it may mean that the rod itself is consumed rather quickly. It is sometimes recommended that the rod end be gently preheated. This is fair enough, but if any length of rod gets hot it loses all its strength and as soon as

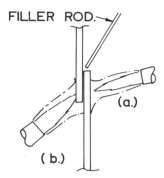

Fig. 49 *Applying the heat to a lap joint. 'b' is the preferred method.*

you apply it to the work it bends and the last state is worse that the first. There is also the risk that you may melt the rod inadvertently and it will run into a blob. The answer to the difficulty is to bring the work, just at the start of the run, to be really hot. This will do no harm so long as you don't get it so hot that you melt it (of course, you wouldn't be using brass wire or spelter ON brass!) and the heat transfer is increased. As soon as the rod does melt, the heat transfer from the liquid in the joint to the rod end is considerably improved and no trouble should be experienced there-after. But on no account envelop both joint AND rod in the flame, for if you do the rod *will* start melting and you will lose control.

I have spent a little time on this be-cause 'brass-wire brazing' is nowadays a sadly neglected craft. It has its place, especially on steel (even more so on wrought iron) and is cheap, for almost any brass wire will serve. If high temperature flux is used (I use the same as I use for Sifbronzing) little diffi-culty will be found, the only point being that the work must be really hot – 900°C

Fig. 50 *A typical application of silver-brazing/flux 'paint'.*

is very much lower. Indeed, for the 'Easy-flowing' types AG1 and AG2 the work will be barely red heat at all. The behaviour of the flux is better, and the alloy is far more fluid at the operating temperature. A final advantage is that the flux is more transparent when at the working temperature and you can see better what you are doing.

The procedure is as before. Initial cleaning, then the application of the flux, within the joint gap and over it. Flux the rod end(s). Apply the heat gently to dry out the flux and then 'turn up the gas'. The flux may discolour a bit, especially if you direct the flame right at it; far better to heat indirectly. As it melts, keep an eye open for any spots which may be short of flux and apply some more with the end of the rod. (Dry flux powder, that is, not the wet paste.) Then, as you heat the end of the joint you will notice that the flux changes its appearance in a marked fashion. I find this very difficult to des-cribe – instead of lying as a simple pool of molten material it begins to 'look active'. This means that you are almost AT the brazing temperature and a little more heat will do the trick. You may suffer the same reluctance of the rod to melt, but this is nowhere near so pro-nounced as it is with the brass wire, and as soon as it DOES melt it will appear to flash round the joint. That is, provided all is hot enough. Once you have the silvery stripe that indicates molten alloy within the joint carry the flame forward with the rod applied a little way behind and you will be surprised how quickly the joint is made. As before, flame ap-plication should be as shown in Fig. 49 (Where appropriate).

A slight change in technique is some-times needed when using silver-brazing alloys having a wider melting range.

or more, and that is at least the 'bright red' of the hardening colours.

One final point – and this applies to ALL brazing, not just to brass wire work. You *must* give the alloy time to flow through the capillary gap, so do not be in a great hurry to move the flame along the joint. When using the silver-copper alloys the penetration is pretty fast and can usually be seen at the back of the joint, but even so it does take measur-able time. The golden rule is to get up to temperature quickly (once the flux hass settled down) get up to an adequate temperature, apply the heat to where the alloy is to flow rather than to where it is to flow *from,* and hold the heat long enough for the capillary to be filled.

Silver Brazing Alloys There is very little difference in the actual procedure when using silver-bearing brazing alloys other than the fact that the temperature

74

These are not so fluid and, as we have seen, can be subject to 'liquation', and whereas AG1 or AG2 (and their cadmium-free counterparts) will usually flash right round a joint of reasonable size from one spot these less fluid alloys will not. This means that you may have to apply the rod at successive spots right round the joint (taking the heat along as well, of course) in order to be sure that the capillary gap is completely filled and that the fillet is uniform. In this respect the procedure is exactly the same as when using 'spelter' though the temperature is much lower.

This sounds very easy, and so it is. Indeed, for many fabrications I find it far easier to braze up with AG2 than to use soft solder. In fact, the actual brazing *technique*, once the simple rules have been understood, is perhaps less difficult than the design of the joint and the selection of the appropriate alloy and flux. Try it and see! Make some practice pieces like Figs. 22 and 49.

Brazing Paste, or 'Silver Solder Paint'
This is, as already explained, an intimate mixture of fine alloy dust and flux, usually in an organic carrier liquid to make it of usable consistency. It is deceptively similar to the soft-solder paint already discussed, but its use is quite different. It is NOT a tinning expedient; though one can sometimes 'tin' a silver-alloy brazed joint with alloy (*not* tin, of course) before the joint is assembled this is seldom necessary. Brazing paste is intended for brazing up very delicate joints where it would be impossible or very difficult to apply a rod, and in industry it finds its main use when many articles are brazed at once in a muffle furnace. For the model engineer and the amateur generally it should be reserved for use on very small articles, of which Fig. 50 is an example.

As received the paint holds the correct proportions of flux and alloy, but it can be thinned a little if desired. Most makers suggest methylated spirit but I find that butyl alcohol (obtainable from any good chemist) is better and does not evaporate so quickly. The same substance can be used if the paint 'goes thick in the jar'. The paste should be applied as a 'blob' at the joint and then heated indirectly if at all possible. Direct heating with a flame usually causes the paint to melt before the work is up to temperature and an unsightly joint is the result. If direct heating IS the only way open, then every care must be taken to ensure that the work heats up before the paint. One way of dealing with this problem is to set the workpiece inside a copper tube and apply the flame outside the tube; the work will then be heated by radiation, and can be left to cool down without disturbing anything.

You will suffer nothing but disappointment if you try to make joints by painting surfaces with the material, within the capillary of the joint. It is definitely NOT appropriate for the 'sweated' sort of joint used with soft solder. Certain brands of solder paint are supplied with a very free-running flux so that they can be applied from a hypodermic syringe. These types are excellent for their purpose, but if used other than for the sort of work already mentioned you must remember the force of gravity! You may find that flux and alloy run *away* from the joint before the alloy melts. Where possible such material should be used on joints so situated that both gravity and capillary attraction will draw the alloy into the joint.

The one area where I find this type of alloy/flux combination ideal is where light parts must be held together by hand simply because no jig can be contrived for them. If a fixed torch is used (see Chapter 9) and a solder-blob applied to both parts, these can be held together by hand, the flame applied to the heavier of the two parts, and then carefully moved away from the flame to cool. The brazing clamps shown in Fig. 68, page 97, are invaluable for this class of work.

Step Brazing Sometimes called 'Sequence Brazing', in which further parts must be brazed onto a component already brazed up. For reasons already mentioned it is sometimes quite pos-sible to braze 'in sequence' using the same alloy throughout. If the initial joints have been held together by rivets or screws, or are self-locking in position, then it doesn't matter a great deal if the alloy in the previous joint does melt — PROVIDED IT HAS BEEN FLUXED. Moreover, the alloy will, within the joint, have combined with the parent metal to some extent and will have a higher melting-point anyway. I have frequently made as many as three successive brazing operations using the same alloy, and even more often using AG1 followed by AG2, where the solidus of the one is the same as the liquidus of the other.

When faced with this sort of fabrication the best thing to do first is to look at the design to see if step brazing can be avoided. It may be that parts can be held together by screws — temporary, or to be brazed in and filed off — or that some form of interlocking of parts can be devised to prevent movement during the making of the later joints. It is not wise to rely on clamps if the work is to cool down between heats, as thermal expansion often causes these to work loose. However, all too often a sequence braze cannot be avoided. Fig. 51 shows the lower section of the boiler already seen in Fig. 23. Here it was necessary to machine slots to receive the bearing brackets so that they could be located within fairly close limits. These were then clamped up and brazed in place, after which a further machining operation had to be carried out on the bracket faces themselves. This could not be done with the boiler 'all in one' with my limited facilities, which meant that I could not braze together the rest of the boiler at the same time as the brackets. There being no way out of this predicament, no less than five steps

Fig. 51 *A typical step-brazed piece, set up on the rotary table. This is the lower part of Fig. 23.*

were involved. The procedure is described in detail on page 113.

As I have said, it is not essential that the solidus of the first alloy used lies below the liquidus temperature of the second, and so on down the line. Years ago this was always thought to be imperative – in model engineering circles – and as a result initial steps often had to be made with an alloy (older readers will remember 'B6'!) with a liquidus around 830°C. I would go as far as this only if the second or third stage was going to involve a long heat with some handling of the hot workpiece, likely to disturb the previous joints. (It is worth going to some trouble in designing the joint, so that the length of the heat is greatest with the first steps and least with the last.) After you have had some experience with sequence brazing you will find that a large gap between the liquidus and solidus temperatures becomes less important. However, for those approaching this type of work for the first time it is better to have something in hand.

Fig. 52 shows the sort of chart I use; this displays the melting range of alloys in BS 1845/1984, with a few from the 1977 edition as well. I have left out the 'specialist' alloys. On the left are the cadmium-bearing alloys; in the next batch are the tin-bearing; next are the straight cadmium-free and finally, for the old stagers, I show the alloys which they will remember by old names. You will be able to see at once that all the cadmium bearing alloys have such low solidus temperatures that no 'gap' is

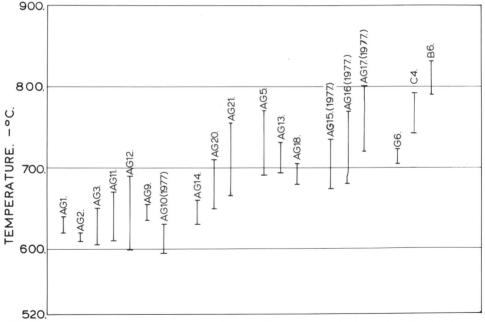

Fig. 52 *Specimen selection chart for alloys, showing the solidus and liquidus temperatures. This is drawn up from BS.1845; it is better to use the actual manufacturer's figures.*

available – though AG2 can follow AG1 if care is taken. There is no gap, but no overlap either (but see later note about actual solidus/liquidus temperatures). AG9 offers a gap, but this alloy is a 'special application' material. However, moving over to the tin-bearing section, AG14 not only offers a reasonable temperature interval but also has a short melting range. So, AG14 followed by AG2 makes a 'matched pair' for two step work – the more so as AG14 is a very nice alloy to work with. For three steps, AG14 COULD be preceded by AG21, but this does have rather a long melting range so that 'liquation' could be a problem. (The fact that it HAS a long melting range means that the overlap of three of four degrees will not be important.) However, the 60% silver alloy AG13 displays both a large gap and a short melting range and this is a reasonable choice for the first of three steps.

Fig. 52 is not intended as a 'directive' – just to show you how to go about things. Most manufacturers of brazing alloy can offer materials to DIN or US specifications as well as to B.S.I., and some list a special range particularly devised for step brazing. All you have to do is to make up a chart like Fig. 52 from their catalogues. And here you will find that the position is much more favourable than Fig. 52 suggests, for this is drawn up using the *midrange* temperatures of the tolerance permitted in BS 1845, The makers catalogue will give the ACTUAL figures; for example, BS quotes 620/640 for AG1 and 610/620 for AG2. One maker offers alloys to these specifications where the temperatures are 620/630 and 608/617, rather more favourable. Use of the maker's catalogue data also helps in the all-important matter of capillary flow char-

acteristics. On the face of it the nickel-bearing AG9 might follow AG2, but it has *very* poor flow characteristics and it is, in fact, intended for brazing carbide tool-tips using a preset foil within the joint. In the last analysis, of course, the choice may be limited by 'what you have' or 'what you can get'!

The actual procedure in step brazing is no different from normal, but you must, of course, use a high-temperature flux in the early stages. I always lay on a layer of flux on the edges which will be brazed later, to keep them oxide free as far as possible, and when working in copper I always pickle the assembly so far done in 5% sulphuric acid to remove all oxide from the rest of the job as well as to remove the last traces of flux. To be on the safe side, and especially when using alloys with an overlap or close gap between solidus of one and liquidus of the other, I always flux the joints already made.

Heating problems To deal with one already mentioned – that of *liquation*, where the eutectic content of the rod melts and leaves behind a higher melting point residue. It is quite obvious that you must have the work a fair bit hotter than the listed liquidus, so that the heat will travel from the work to the rod very quickly indeed. Normally one will be using these long melting range alloys on largeish work involving sequence brazing. Model loco boilers are typical examples and here the usual rubric is to heap coke around the parts of the boiler shell which both acts as a heat insulator and also provides some heat as it burns. This is quite sound provided the remarks made previously about dirt adhering to the next joint are borne in mind. I use heat retaining chippings, obtainable from William Allday & Co.

Ltd, Stourport on Severn DY13 9AP, for the same purpose. However, if you have the facilities a better method is to use a second torch. The main brazing torch is used with a hot, cone-shaped flame to play on the actual joint whilst the auxiliary is set to give a broad blowlamp type flame, played on the main body of the work to keep it hot generally. Both can be hand held at first and then, when up to heat for actual brazing the auxiliary torch is set on a bracket or clampstand to give an envelope of flame as far as possible over the body of the work (see Fig. 54). The old Alcosa brazing hearths had adjustable clamps for two auxiliary torches and they were better still – but rather expensive these days. This twin-heat method does enable the actual joint to be brought up really hot and so reduce the liquation difficulty.

Another trick is to apply a very small amount of short melting range alloy right at the beginning of the joint – just a touch of (say) AG2, and then *immediately* contact this molten blob with the end of the true brazing rod. The heat transfer through the molten metal is very much faster and with any reasonable luck the high-range alloy will really flow. It is not always easy to do this – one needs three hands – but you may be able to pick up a tiny fragment of AG2 (there are always little unusable ends left from previous jobs) on the fluxed end of the main rod. This does work, though there is a tendency for the fragment to drop off at the critical moment! The very small quantity of AG2 will have no deleterious effect on the joint as it will be 'incorporated' in the other alloy straight away.

Distortion is another difficulty, especially when brazing long joints in thin-wall material. This is due to two causes.

First, if cold-rolled sheet is being used there are locked up stresses which are released as the metal approaches red heat. The cure here is to anneal the raw material thoroughly before starting to braze. The second cause is differential heating. The joint line is hot and the rest much cooler, so that the metal tends to buckle. This is very difficult to avoid and the only solution is to take the bull by the horns and make sure that the *whole* of the work is at a uniform temperature, and heated uniformly in the process, too. A 'similar but different' problem can arise when relatively light gauge material must be brazed to a heavier section. The rule here is quite definite; the heavy section must be heated first and not until it is at the brazing temperature is heat applied to the lighter section. Indeed, it may well be that this will come up to temperature simply by radiation from the rest of the work. AMPLE flux must be applied in such cases, as the heating may be prolonged and the flux exhausted before the larger mass of metal is hot enough. The same applies when brazing two *different* metals; heat the one which is the worst conductor of heat first.

Slow working. This is by far the commonest fault in amateur work and not uncommon in professional as well. Unlike heat treatment, where slow heating is the rule, heating for brazing should be rapid once the flux has settled down and especially so between the actual melting of the flux and the application of the rod. It is unfortunate that the fluxes which have the best fluidity and the highest activity have a relatively short life, whilst those which have a long 'active period' are LESS active and may have poor penetrating power. This is especially true of those in the upper temperature ranges. I shall be saying more about heating torches later, but it really is essential to be sure that the heating method is man enough for the job. If in doubt fire up the old paraffin blowlamp and use this as an auxiliary burner. (No need to fiddle with meths and so on – just heat the burner coil with the main blowlamp and then give a few strokes on the pump!) Flame heating is very inefficient, especially when the work is largeish. It is useless to keep 'struggling on' hoping to get up to temperature eventually; the end result will be a bad joint and all to do over again. In addition, never be hesitant about adding flux (in powder form). You can do this using the brazing rod, the end of an old hacksaw, or even use a long-handled spoon! It is no good at all trying to proceed once the flux is exhausted – the alloy won't wet the parent metal, won't penetrate the gap, and won't make a bond even if you persuade it to form a fillet.

Overheating is less of a problem, but when working in brass it is as well to keep an eye on things. In Fig. 51 the little brass bearing pedestals have a very small heat capacity compared with the body of the work and if a too powerful flame is used it would be possible to melt them. The trick, again, is to use a broad, soft flame to heat the body and then apply the second, brazing flame to the joints once the whole is almost up to temperature, but to keep the brass parts out of flame until then. However, there is one circumstance when serious overheating can occur, and that is when using oxygen/fuel gas blowpipes. The temperature at the cone of these flames is VERY much higher than in any air/gas burner, and it is all too easy to melt the workpiece – the torch is, after all, intended for that purpose! Any attempt to braze over such an area will fail; even if

it 'looks alright' the bond will be on to burnt metal and be very weak. But far more serious is the risk to the brazing alloy itself. If the hot cone of the flame meets molten alloy cadmium boil-out is certain, and zinc boil-out probable. Both circumstances are VERY hazardous to health (they can be fatal) and disastrous so far as making a good joint is concerned. Nor will the flux stand up to such high temperatures, either. As I shall point out later, oxy-gas torches have their uses, and in experienced hands are quite satisfactory (experienced brazier's hands, I mean; welders unaccustomed to brazing will almost certainly overheat the job) but though I have used 'Oxo' for 40 years, and still do for special work, I use gas-air torches for almost all work.

Slag and Flux Inclusions These are as much a fault in joint design as in technique; if the alloy is to penetrate the joint it must be able to displace the flux which is protecting the surface prior to its entry. And if the flux cannot get out, where is it to go? Apart from preventing access by the alloy it forms a reactive chemical inclusion which will very likely cause a later failure. Further, the alloy cannot travel through the capillary if it 'freezes' on the way; the place for the hot part of the flame is at the *back* of the joint, where the alloy is to go TO, not at the front where it is coming FROM. The majority of the 'pinholes' and leaks experienced by model engineers are the result of flux inclusions, but inadequate initial cleaning can result in a wetting failure by the alloy.

Brazing Stainless Steel The term 'Stainless Steel' covers scores, if not hundreds, of alloy steels, and brazing alloys which will serve for some will be quite unsuitable for others. Thus AG9 will serve for most Austenitic (so-called '18-8') steels but not for the Ferritic types. It is imperative that the alloy maker's advice be sought for any important joints. (For plain water-tanks etc, soft soldering should be quite adequate.) The brazing alloy can then be tailored to the base metal, and to the intended service of the joint. This is especially important when considering superheater joints. You will find that the manufacturers are quite happy to advise, as a joint failure in such an important situation is a matter for concern for all. However, do be reasonable! Don't expect a firm to engage in lengthy correspondence over a job which may use perhaps 10 grams of alloy! Make your enquiry brief and to the point, and you will get a careful reply.

The very nature of stainless steel makes the choice of flux equally important. Fortunately proper grades of flux are readily available, but it is wise to avoid prolonged heating. A general recommendation when brazing drawn tube or cold-rolled sheet is that the stock should be stress-relieved before brazing. Remember that stainless steel is a poor conductor of heat and that most of the special brazing alloys have poor fluidity, so that wide lap joints can present difficulties. There should be no need for wide laps, as the shear strength of the brazing alloys is quite high and in any case these alloys all tend to form fillets.

Having said all that I must confess that I have made scores of joints for small components in stainless steel using nothing but AG1 or AG2 and 'ordinary' flux. The few failures – usually refusal of the rod to wet the parent metal – have been traced to the fact that one of the parts was of a 'stainless' steel of unknown origin and composition. It is the parts which are subject to fatigue stress, high (up to 200°C) temperatures, or which are to spend most of their time in mucky water that need 'care and attention' in the brazing.

Brazing Tool-tips Both high-speed steel and tungsten carbide tools can be brazed without materially affecting their performance at the cutting edge as the brazing temperature is well below the nominal 'temper' of the material. It is seldom that model engineers or amateurs can achieve the cutting speeds necessary for the economic use of carbide, but the ever-increasing cost of HSS does make the composite tool, with HSS tip brazed to medium or low carbon steel shank, worth looking at. Again, though soft solder is often adequate when making an extension to a twist drill, brazing is more secure and if fine joint gaps are used alignment is better. With the small size of tool normally used there is no need to employ the special alloys (AG9 etc) and as long as the tip is not much above half-inch wide the standard AG1 alloy will usually flow right across using normal flux.

The tool tip, whether carbide or HSS, should be lightly ground so that the grinding marks run in the direction in which the alloy will flow, and the shank should, if possible be treated the same. Both should be thoroughly degreased – this is very important, as it is vital that the flux makes good contact. The tip should be bound with black iron wire to hold it in place after fluxing both shank and tip, and then heated from the shank. The alloy is applied at one side and can be seen to have 'arrived' at the opposite side by the appearance of a fine silvery line. A further application can be made at the tip if desired, but this should not be necessary. I always do this job with the tool upside down (tip below the shank) with the overlap of the tip forming a little shelf on which to apply the rod. Some authorities recommend quenching the work from black hot to remove the flux, but I prefer to aircool right down and then wash off flux in hot water – I feel that quenching does present a risk of cracking the tip. Care should be taken that the metal is hot enough to ensure good penetration of the alloy but remember that the flux has a limit to its life. The joint faces must be a good fit, with an average joint gap not exceeding 0.003 inch.

Cast Iron Cast iron can be brazed using AG1 or AG2 and normal flux provided that the surface is newly machined or filed. However, any trace of oil on the surface does seem to cause a lot of trouble, and this is probably due to the free graphite in the iron absorbing the oil. (For 'stressed' joints the alloy AG18 is recommended.) As 'bronze-welding' (Sifbronze) was first introduced for the repair of iron castings it may be that higher temperature alloys do not suffer in the same way, but I confess that though I use bronze-welding frequently I have never tried simple *spelter* brazing of iron other than with the 'easyflowing' alloys. The reason is that the heating of anything but a simple 'lump' of cast iron must be done with some care. For example, if a crank-boss is to be brazed onto a C.I. flywheel any rapid heating of the casting could well cause the spokes to crack. At the low temperature needed for silver brazing this is easy to manage, but for spelter brazing a fair time for preheat must be allowed, and the flux might well be exhausted before higher temperature brazing started. Brazing at around 620°C presents no problems,

but plenty of flux must be used, renewal during any preheat may be necessary, and care must be taken that the casting is neither heated nor cooled unevenly if cracking is to be avoided. For simple jobs, like attaching a (model) flange to a pipe, or a chucking piece to a simple casting – or even a simple fabrication in iron – you should have little trouble provided the parts are 'clean metal' to start with. Shot-blasting might be the best answer for those who have the equipment.

Bronze-welding lies outside the province of this book but, in passing, it is worth noting that it may well be the answer to awkward repair jobs in cast iron in those cases where welding is not practicable. The process itself is simple enough, but attention must be paid to the preheating of the work.

Conclusion Brazing is no more than cleaning the pieces, applying the flux, assembling the parts so that they cannot move, applying the heat and bringing the joint up to the right temperature at the right place, applying the alloy, and removing the flux after cooling. It IS as simple as that and provided that the joint is properly designed, as easy as, if not easier, than making a good joint with soft solder.

A MOST IMPORTANT NOTE I have left this till last, so that you will notice it. Especially when working with shapes which have been beaten out, but fairly frequently even with machined joints, you may find that there is a joint gap well beyond the maximum permissible. If this can be closed by further work with the hammer, well and good, but if not, then you MUST fill it with a packing-piece of the same material as the main body of the work, so that the alloy is not asked to bridge a space larger than the normal capillary flow area. I keep a little box containing scrap offcuts of copper etc from which I can cut little wedge-shaped pieces for this purpose. These can be tapped in place and the wedge-shape will retain them whilst still filling the hole and leaving a reasonable gap either side for the alloy to penetrate.

Chapter 9

Brazing Equipment

The Brazing Hearth Unlike soft-soldering, which can be carried out almost anywhere, the high temperatures involved do require some special provision when brazing. And here let me proffer a word of AWFUL WARNING. If the printer could manage letters of fire I would ask him to do so! **ANY CONTAMINATION OF A BRAZED JOINT WITH SOFT SOLDER WILL BE FATAL.** True, the brazing alloy will 'run' – perhaps run very well – but the joint will snap off like a carrot if any soft solder is present. So, never braze over soft solder and above all, never soft solder in the brazing hearth, lest some unnoticed splash wreak havoc with the joints.

I have already mentioned that brazing can be done in the blacksmith's hearth or over charcoal, but most practitioners today will be using a gas torch or blowlamp of some sort. You can buy proper hearths for such use, almost always with the great convenience of a rotating table, but these are costly and unless you are going to be brazing frequently a home-made affair will be quite adequate. Fig. 53 shows my smaller brazing spot, made up entirely of commercial firebricks. The base is a 12 x 12 x 2 inch slab, and this is set on top of four 'Fossalcil' heat insulating bricks for safety. There are more standard firebricks at the back and one end, the other (RH) end being one of the bricks sold to reduce the volume of a domestic fireplace. Immediately behind is a sheet of steel with a sheet of soft asbestos board in front, to ensure that stray flame from the torch does not harm what lies behind. It is better if this sheet of steel curves over forwards just a little, as even on small jobs quite a lot of heat travels upwards. (The ceiling lies a good six feet above the hearth.) This arrangement serves for all my casual fabrication work – anything that can be handled with a flame about the size of that from a 1-pint paraffin blowlamp. It also serves as my tool-hardening hearth, and the electric muffle is just outside the picture on the right. But it is too small for any major work and anything needing a flame more than about 6 inches long is done outside the workshop.

Fig. 54 shows the larger hearth. The 'bench' is timber, which is not really advisable. However, there is a sheet of flat fireproof material (the current replacement for hard asbestos) on top,

Fig. 53 *The author's 'small' brazing hearth, used for odd jobs which need only a small blowlamp.*

and a number of Fossalcil heat insulating bricks over this supporting a 20 gauge steel tray, 22 x 34 inch long. This latter is filled with ordinary 9 x 4½ x 3 inch firebricks bedded in sand to form the working surface, but at one point, on the left, a gap is left which accepts the small rotating table about a foot in diameter. This is a commercial one (supplied by Flamefast Ltd) but I used to have one made up from an old lorry rear brake drum, filled with firebrick. You will see firebricks all around the ends and back. Some are ordinary ones, some Fossalcil, and some the 'Hot Face' bricks now available (see later). These serve first to protect the walls and second as a store of bricks for supporting work. The walls themselves are masonry but, nevertheless, protected

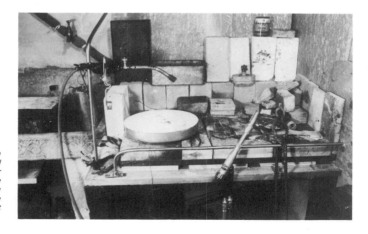

Fig. 54 *The author's main 'brazing centre', with small rotary table, clamping pillar for fixed torches, torch rest at the front, and a stock of insulating and firebricks around. Note the small paraffin lamp used to 'light the gas'.*

Fig. 55 *The Alcosa rotary brazing hearth with torch and blower. (Courtesy William Allday & Co., Ltd)*

against casual flame impingement by flat sheets of hard asbestos.

At the front of the hearth is a rail – a guardrail from a scrapped corporation bus – on which torches and tongs can be hung, and at the one end a similar piece of material fitted to a socket on the base to which a torch can be clamped. The shape is a bit odd, but it is truly 'universal' and the torch can be set to point in any direction and at any angle. This means that I can have one torch hand held, another clamped to play over the main body of the work (the clamps are such that the outfit CAN be moved, but won't move by itself) and still leave one hand free to manipulate the rod. The fuel cylinders live under the bench, as does the air-blower (and my workshop compressor) and there is also a shelf underneath for odds and ends. Conveniently, there is a sink to the left-hand side in which my pickling-bath lives. In this case the roof (there is no ceiling) is about 15 feet above the hearth, so that there is no need for any overhead protection.

This is a point to bear in mind. With my biggest torch going full blast the amount of air in the shop would reach 'temper blue' in about 20 minutes! and the hot air rises, so that if there IS a ceiling above it may need protection. Imperative, however, is proper ventilation. In my case there is a door to the outside alongside the sink and when using any but the very smallest torch this is open. Comfort demands this even in the depths of winter, but equally, so

Fig. 56 *A larger Alcosa brazing bench, fitted with a rotary table. The blower for the torch is set within the casing. Note the gas control cocks. (Courtesy W. Allday & Co., Ltd)*

does health, especially when using cadmium-bearing alloys. I shall have more to say about this in the last chapter.

These illustrations will give you some idea of how to go about things. Your personal convenience is as important a matter as anything else and such matters as bench height and layout will be quite different for different people. However, to give you some idea of commercial devices. Figs. 55 and 56 show two that are available from William Allday & Co., Stourport on Severn.

Now a word about firebricks. The normal type are 'refractory' — they will stand almost any temperature you can apply, but they are not particularly good heat insulators. They can be had in a wide variety of sizes from builder's merchants, from one inch thick up to around three inches and (if you are lucky in your merchant) up to perhaps 18 inches square. They are also available in a wide variety of shapes to accommodate the various types of firegrate. The INSULATING bricks which I have mentioned are NOT refractory; they will only stand up to about 1000°C before melting; they are very light, almost like cork, and fragile. While firebrick needs to be cut with a chisel these bricks can be carved, or sawn with a handsaw. They are normally intended for setting behind the ordinary firebrick, which protects them from the flames. The two which I know are called 'Fossalcil' and 'Folsain' and both are available from good builder's supply houses. I have seen none larger than 4½ x 9 inch, but they can be had in thickness down to 1 inch. They are very good insulators indeed. The third type of brick is a combination — they have good refractory properties and good insulation, too, but not as good as either of the special purpose bricks already mentioned. To the best of my knowledge they are not used by builders and must be ordered from specialist suppliers; William Allday & Sons Ltd, Alcosa Works, Stourport on Severn, Worcestershire DY13 9AP, Messrs Flamefast Ltd, Pendlebury Trading Estate, Manchester M27 1FJ, or from MPK Insulation Ltd, Hythe Works, Colchester CO2 8JU. They are rather expensive, and carriage is not cheap!

Finally I should mention the old fashioned 'Charcoal Block'. This, perhaps 3 x 2 x 1 inch thick, is still used by working silversmiths. The work is laid on the block and when the blowpipe or torch is applied, acts as a secondary source of heat. The charcoal starts to glow under the flame and adds radiant heat, and, of course, burns also, to add a supply of heat from below. The charcoal is consumed only very slowly and a block will last a long time. I use these for all delicate work; to be had from any of the horological or jewellers' supply houses. Incidentally, 'Barbecue Charcoal' can be used to pack round work as a heat insulator and, as it burns, helps to keep the job warm; better than coke as it is cleaner and free from sulphur, but don't forget the ventilation, as slow-burning charcoal can produce poisonous carbon monoxide. I use heat retaining chippings — a special form of ceramic — which don't leave any dust on the work. Obtainable from William Allday and from the Flamefast companies, and usually in 5kg bags.

Heating Methods As we have seen, you can use anything that will 'put the heat in the right place', and some very exotic devices are used in industry. However, it is interesting to observe that flame or torch heating has been classed by production engineers as the most versatile, and least expensive in capital costs, of all forms available when dealing with 'one off' or low volume production, the only qualifying factor being that it needs some 'operator skill'. Such skill is a very expensive commodity in industry, but it costs US nothing – just some time in practice, that is all. So, a flame of one sort or another we shall use, the important question being 'what sort'?

First, a very brief word about fuels, though these are covered in more detail at Appendix II. These are nowhere near as critical as you might imagine so far as the heating power is concerned. In fact, when mixed with the chemically correct quantity of air the heat output per cubic foot of air-fuel mixture is the same – within 1% or so – whatever fuel is used. True, the maximum flame temperature varies quite a lot, but even the worst of the fuels we are likely to use reaches 1900°C, far higher than we need for brazing. So, if you are 'on the gas' in your street this can be used for brazing with a suitable burner. Paraffin, used in the conventional blowlamp, will serve and even methylated spirit, blown with a mouth-blowpipe, will braze up small objects. Any fuel will do – but the *size* of the work you can handle will depend on how much fuel the burner can use – it is as simple as that. To be really precise, it depends on how much AIR the burner can handle, for we could pour out paraffin by the gallon (or liquid gas for that matter).

The limitation which we meet, when considering gas especially, is the *rate of supply*. If on the gas main, using North Sea Gas (methane) you may have to have a word with the engineers at the Gas Board to make sure that your meter will pass sufficient for your needs. A really large gas blow-torch will use possibly three times as much as a large gas cooker running flat out on all burners. Again, when considering 'bottled gas', you may use Calor butane in your caravan, and it seems a bit stupid to have a cylinder of propane as well – after all, you are not likely to be brazing up a boiler when touring in the van. But there is a snag. The gas is liquid, and as you draw off the vapour the latent heat of vaporisation must come from the liquid in the cylinder. So, it gets colder. But whilst propane will still 'boil off' at $-44°F$ ($-42°C$) butane needs at least $+14°F$ ($-10°C$). So, you can take gas from a propane cylinder much faster than you can from a butane bottle. Moreover, if you are using a self-blown torch (one with no air blower) the pressure in a butane cylinder (around 30 lb/sq.in) will be insufficient fully to 'drive' the burner; propane will normally be stored at around 140 lb/sq.in. This doesn't arise with a burner having its independent air supply, when either gas will do just as well (provided the gas demand does not cool the cylinder too much) for such torches require a gas pressure of no more than around 11 inches water-gauge (0.4 lb/sq.in). I shall have more to say about these two types of burner later. The main point to notice is that given the right burner and an adequate supply system any of the common fuels will do all we need.

Paraffin Blowlamps Years ago these were the standard heating method for those not on Town Gas, and this in the days of 'spelter brazing' at that, needing temperatures up to 950°C. The physical size of the larger (five or seven pint capacity) ones made brazing a two-man job, and they fell out of use as soon as 'portable' gas became available. However, for work within its capacity the smaller one-pint blowlamp is not to be disdained. The fuel is cheap and safe, it has other uses both in the workshop and about the house, and the blowlamp itself is relatively cheap, too. The main objection I hear about them is the trouble of lighting them. Preheating with methylated spirit always seems to put people off! However, there must be few households who do not have one of the tiny 'Camping Gaz' type blowlamps (see Fig. 39). It is not man enough for any but the smallest of brazing jobs, but it *can* be used to fire up the heavier paraffin lamp. Just light the gas, heat up the coil inside the paraffin blowlamp's mouth, and within a few seconds you can pump up and the torch flame will emerge.

The more valid objection is that they have a habit of running out of fuel at the critical moment. This means no more than that you have to see that it is full before you start – a fully charged tank should normally last about an hour. The type of flame is well adapted to brazing, and you will be able to do most of the fabrication jobs which arise in general model-making, household repairs, and even small boilers. If you have access to 'Royal Daylight' paraffin (or 'Tractor Vaporising Oil' – TVO) this is better than the rainbow-hued stuff normally sold for paraffin heaters these days, but the latter will perform well enough. A 1-pint paraffin torch is more or less equivalent to a propane burner using

around 12-14 ounces of LPG/hour (or say 7 cu.ft/hr on propane and 16 cu.ft/hr of North Sea gas).

Gasfired Torches These are available in two distinct types; those which need an independent air supply and those which 'blow themselves', rather like the paraffin blowlamp. The former use gas at very low pressure – that delivered from the domestic gas pipe or, at the most,

Fig. 57 *The Flamefast air/gas torch, type T4, in use. (a) set for needle-flame. (b) A full-power high temperature brazing flame. (c) Large soft flame. The change in flame conditions requires only adjustment of the air and gas valves. (Courtesy Rhodes-Flamefast Ltd)*

89

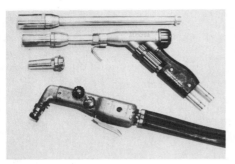

Fig. 58 *The author's air/gas torches. Top. Alcosa 30/70 fitted with medium-size nozzle; their larger (60,000 BTU/hr) nozzle above has an extension tube. Below is a small nozzle for delicate work. Bottom. The Flamemaster precision torch, designed for glass-blower's work, but very useful for fine brazing. It can be used either with air or oxygen.*

12 inch water-gauge (about 0.4 lb/sq.in). The air supply can come from foot-bellows (universally so in the old days). I *have* used my workshop compressor with a reducing valve for this purpose. The burner is of the injector type and the air stream collects the gas and mixes with it before arriving at the

Fig. 59 *The author's Sievert self-blown torch outfit. Top, D.3 handle with flame-guard and large burner. Centre, E.3 handle with medium-small nozzle. Below, Range of nozzles from 90,000 down to 1500 BTU/hr, having various flame characteristics. Bottom, 'Neck-tube' burners for use in confined spaces. (Note: The current range of handles and burners differs from these in detail)*

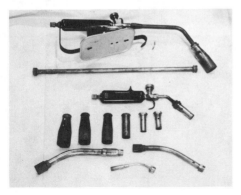

flame tip. The *self-blown* type requires a high gas pressure, of from 28 up to 56 lb/sq.in, but no pressure air supply. In this case the *gas* jet induces the *air* through the ports in the burner nozzle, mixes with it, and again, ignites at the burner mouth. The first type can be used with ANY gas, with a suitable low-pressure regulator on bottled gas, but the self-blown type cannot be used on 'town' (North Sea) gas. Further, because butane ('Calor Gas') is stored at only 30 lb/sq.in, and the pressure drops to about 20lb in the winter, the full duty of the self-blown torch cannot be realised. In general they are limited to use on propane.

Which is best? The short answer is that either will do. You will need an air supply for the air-blown burner but not for the self-blown type. On the other hand, the fact that you can adjust both the air *and* gas supplies means that you can adjust both size and type of flame (see Fig. 57). With the self-blown unit the type of flame is fixed by the burner design — needle, broad, fierce, or soft — so that you have to change burner heads for different jobs. You may need four or more burner heads to duplicate that which can be done with a single nozzle on the air-blown type. The self-blown type is inherently portable — you can use it anywhere, but the other must be within reach of the air-blower. I have used both for very many years, and much prefer the type with separate air supply because it is so flexible — you can alter the size and type of flame at will. The single nozzle fitted to the unit shown at the top of Fig. 58 does almost all my brazing work, and the other two nozzles cover the lot. But I have used the self-blown types shown in Fig. 59 on similar work, though I then need a choice of eleven nozzles. Fig. 58 will run on propane, butane or North Sea gas, Fig. 59 needs propane. So, you takes your choice! I can only add that were I on 'mains gas' I would never consider any but the air-blown type.

To complete this commentary I show Fig. 60, which indicates that the hottest part of the flame is just beyond the centre of its length. But most of the flame is above brazing temperature, so that this is not as important as when heat-treating steel. Temperatures with a properly adjusted air-blown torch will be slightly higher. There is, however, one matter concerning the flame which is of importance. It does depend on air drawn in from its surroundings as well as on that induced by the injector effect or supplied by the blower. The self-blown torch is most affected, and if this type is used in a confined space it will go out. In such brazing work with a self-blown torch (inside a model loco fire-box, for example) the 'Neck-tube' or 'Cyclone' type of burner head should be fitted. The air-blown torch is nowhere near so sensitive, but both will, of course, go out if the waste gases cannot escape from the cavity.

Oxygen-fed burners These include oxy-hydrogen, oxy-acetylene, and oxy-propane. These burners produce a flame which is VERY much hotter than any air-gas flame; the hottest part of an oxy-acetylene flame will approach 3300°C. On the other hand, the actual heat OUTPUT is relatively small. A No. 10 nozzle – about the largest supplied with the ordinary 'DH' welding set, and capable of welding 4mm mild steel plate – has a heat output quite a bit less than that of a one-pint paraffin blow-lamp. These two factors have important consequences.

The normal type of brazing torch will heat up the whole of the work more or less uniformly, and any temperature gradients will be gentle. Even though the actual joint area may be hotter than the rest the difference is not all that great. There is, therefore, little risk of thermal distortion – the work will tend to expand symmetrically. The oxy-acetylene flame is not man enough to heat the whole of a large workpiece, but relies on the high temperature of the flame to bring local areas up to brazing temperature without heating the rest. The temperature gradient is much steeper and the risk of distortion correspondingly increased. Most people realise this and take precautions, but there is one aspect of the matter which is seldom mentioned. Brazing alloys, and especially the easy-flowing cadmium type, do need small joint gaps – say between 0.002 and 0.004 inch. Smaller than this, the alloy cannot penetrate; larger, and it will not enter the capillary. By its very nature an oxygen-supported flame brings just a short length of the joint up to brazing temperature with the point opposite almost cold. This inevi-

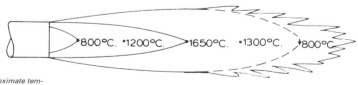

Fig. 60 *Approximate temperatures within an air-gas flame.*

800°C. •1200°C. 1650°C. •1300°C. 800°C.

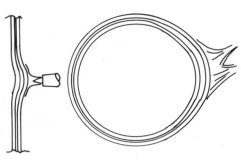

Fig. 61 *Showing how the application of intense 'local heat' can distort the joint gap. The effect is exaggerated for clarity.*

tably causes drastic changes in the joint gap if care is not taken (see Fig. 61).

The second, and very important, point is that if the flame impinges on the alloy itself, especially when molten, cadmium boil-out is certain and there is grave risk of volatilising some of the zinc as well. Quite apart from the risk to the operator — the former can be fatal — the net result will be a bad joint with poor bonding and slag inclusions as well. On the same point, it is obvious that if the very hot cone impinges on the work itself there is risk of melting the base metal too, with even more serious consequences.

In the hands of an experienced opertor the oxygen-bearing flame has advantages; not least that as the actual flame is small it is possible to heat 'the back of the joint' to encourage capillary penetration, and small areas on a large workpiece are very easily dealt with. But the use of this equipment does need experience and I would not advise any beginner to invest in oxy-acetylene, or oxy-propane, equipment until he has done a fair bit of brazing using conventional flames — and even then it would be wise to seek advice from an expert

practitioner first. In which connection I mean an expert practitioner in Oxy-BRAZING; the techniques used in welding are totally different, and can only lead to disaster when brazing.

The one case where oxy-acetylene might be *recommended* is when using the phosphorus/copper alloys. The wide melting-range coupled with the high conductivity of the parent metal (copper) makes the high flame temperature advantageous and there is no cadmium nor zinc present.

Carbon Arc Brazing The 'flame' temperature in this case is even higher than that of the oxy-acetylene burner, the 'flame' in this case being hot gas ejected from the arc region, the carbon being impregnated with a chemical to achieve this effect. Again, the heat output is relatively small, a 2KVA arc producing rather less than half the heat of a 1-pint paraffin blowlamp. But the heating effect is intense and copper-zinc brazing of steel can be done with this type of equipment. I would NOT recommend it for any silver-bearing alloy, the boiling point of silver being only 1955°C! The problem with this equipment is that the wearing of a protective mask is absolutely essential, and the glass is so dark and of such a colour that it is impossible to judge the temperature of the metal. The 'arc flame' too, has a pronounced 'driving effect' and can shift the molten alloy away from the joint if care is not taken. There is no advantage over any of the other methods of heating but obviously, if any arc-*welding* is done in the workshop the addition of carbon-arc equipment will extend its use, though I use mine only when brazing on steel. For the 'do-it-yourself' enthusiast who includes arc welding in his repertoire the system has some limited application, but for true engineering work (model or otherwise) the air-gas brazing torch will give far better results. If you do have occasion to use carbon-arc apply plenty of flux, keep the 'flame' away from the work when applying the rod, and keep the rod well fluxed too. I use prefluxed *bronze-welding* rod as a simple brazing alloy on the odd ocasion when circumstances make carbon-arc necessary.

Micro-brazing Torches This is the only name I can think of for the tiny burners which have become available in the last ten years or so. Intended in the first instance for the jeweller, silver and goldsmith and allied trades, they do have their applications to other work — indeed, anything that is really small. Fig. 50, page 74 is a typical example of the sort of work for which these small flames are ideal, and as my particular brand of model engineering calls for a lot of small detail I make considerable use of such.

Butane/Air I have already shown the small Flameset torch, in Figs. 37 and 47. This uses the Taymar disposable butane cylinder as fuel supply and most ingeniously, a small air-compressor weighing less than half a pound, consuming only 4½ watts of electricity for the air supply. (Actually, one of the little pumps for ventilating fishtanks.) The torch has control of both gas and air and can be either held in the hand or used on a bench stand as a fixed torch. A bit tricky to light at first (the temptation to turn up the gas too high must be resisted!) but an admirable tool once this is mastered.

Butane/Nitrous Oxide The Microflame torch, shown in Fig. 62, uses tiny cylinders of these two gases. The flame

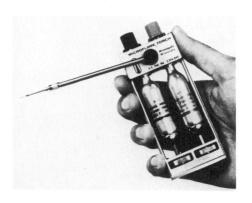

Fig. 62 *The Microflame micro-brazing torch, fuelled with Butane and Nitrous Oxide. (Courtesy Microflame Ltd)*

characteristic is identical to that from oxy-butane, with a temperature of about 2800°C at the tip, but the outfit also includes a low-temperature burner for soft soldering.

The same torch unit can be used with bottled oxygen and butane using an adaptor set (see Fig. 63), and this is advisable if a great deal of work is to be done, for the little nitrous oxide 'cylinders' last about 15 minutes (the butane twice as long). For what it is worth, the torch will fold up and go in your pocket.

Butane/Oxygen The bottled oxygen/butane adaptor mentioned above is normally used with the Microflame Weldmaster brazing torch (see Fig. 63). This is a miniature welding torch and with normal nozzle will handle sheet material up to 24 gauge (0.5mm) thick. (Larger nozzles are available.) The fuel gas cylinder contains ¾lb of butane, and the oxygen cylinder 12 litres. The torch is of the usual twin-valve type, giving control both of the size and type of flame. Had I known of this one when I

93

Fig. 63 *Top, the Weldmaster micro-brazing torch, for use with Oxygen and any fuel gas. (Courtesy Microflame Ltd) Above, the Butane and Oxygen 'Gas Pack' which can be used with either of the torches in Figs. 62 or above. (Courtesy Microflame Ltd)*

bought the nitrous oxide outfit I would have used it instead, but as I have both I find that both have their uses!

It is important to appreciate that all of these are quite unsuitable for anything BUT 'micro' work, delicate detail, and small components. They CAN be used to deal with awkward jobs on larger

work, though. I have dealt with a nasty pinhole in a very difficult spot by bringing the work *almost* to brazing temperature with a normal torch and then 'hotting up' just the offending spot with the Microflame. So long as the size limitations are appreciated these little fellows will give good service, but they CANNOT be driven to heat work which is too large for them.

Gas Generators I include these for completeness, even though there will be few model engineers who will have sufficient detail work to make the financial outlay worth while. The JME Microwelder, by Johnson Matthey Ltd, South Way, Wembley HA9 0HW, generates oxygen and hydrogen by electrolysis of water, and needs only a lead and plug to the electricity supply. Gas is available within a few seconds and the whole process is automatic. The oxyhydrogen flame is very hot (*c.* 3500°C) but of low heat content. However, by passing the gases through what I can only call a 'fuel liquid' the temperature is reduced but the heat content/litre is increased; more heat at a lower temperature. Thus methylated spirit gives a flame temperature about 2500°C, and a special fluid (normally supplied with the outfit) gives 1850°C and very much greater heat content per litre. The torch is very similar to that used on the Microflame with but a single control valve. I have used the 'A' type machine for light brazing and found no difficulties within its capacity — the largest flame tip was 20-gauge.

The Oxyflame, by Williams Research & Development Ltd, Busgrove Lane, Stoke Row, Henley on Thames, Oxon RG9 5QB, works on a different principle and needs no electrical supply. Oxygen is produced by the controlled dissociation of hydrogen peroxide over a catalyst, the residue being plain water. The fuel is a mixture of butane and propane, supplied from one of the usual small gas bottles. Again, the torch and its behaviour is very similar to that used by the other 'micro' outfits, but with a range of nozzles giving flames from less than $\frac{1}{16}$ inch up to 7 or 8 inches long. The gas supply is automatically regulated by the demand from the burner.

General Points on Burners A REGULATOR is needed on all bottled gas (LPG) cylinders. Low pressure (11-14 inch water gauge) for air-blown torches and high-pressure (preferably adjustable, from 14 to 56 lb/sq.in) for propane with the self-blown type. These high-pressure regulators MUST incorporate a hose-failure device, which will shut off the supply automatically should a hose burst. Low pressure supplies do not need this. In all cases you must ensure that the regulator is adequately sized for the gas consumption of the largest nozzle you are likely to use – and think ahead on this matter! The local bottled gas suppliers should have data sheets available, but if in doubt write to Calor Gas Ltd, Windsor Road, Slough, Bucks. SL1 2EQ or to Primus Sievert Ltd,9 Gleneldon Road, London SW16 2AU. Appendix II gives more guidance on this matter.

Note that butane and propane regulators are *not* interchangeable. You need a different one for each gas. If you wish to operate both low and high pressure service from the same bottle you will need a two way adaptor WITH TWO STOP-VALVES; it is most unsafe to rely on the torch valves alone. If two torches of the *same* type are to be used then a simple two-way fitting on the torch side of the regulator will be adequate. I give

some data on gases in the Appendix, and this will enable you to determine the size of blower required for air-blown torches. 5 lb/sq.in is adequate and a good deal of brazing can be done on half that pressure. If you intend to rely on air drawn from the storage tank of your workshop compressor then you really need a proper reducing valve – it is not wise to try to control the supply with a simple screw-down valve.

Vacuum cleaners have been rigged for air supply (Fig. 64), though a motor speed control is desirable to regulate the air flow.

When using large nozzles, an extension tube is essential. The heat radiated from a large workpiece like a 3½ inch gauge loco boiler is considerable. A simple heat shield can be rigged up on the handle (this is standard on the larger Sievert torch) and this helps a great deal. A proper heat-resisting

Fig. 64 *A vacuum cleaner can provide air for an air-blown torch; this one is the Alcosa 3080 (Photograph L. M. Wade).*

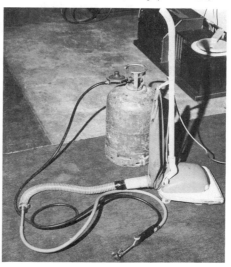

glove on one hand is worth while, too, but make sure that it IS flameproof, for a glove on fire can be an embarrassment.

On burner selection, you must use your judgement. There is little problem with the air-blown torch, as the 'turn down' ratio is very good, and one burning about 12 cu.ft/hr of propane or 30 cu.ft/hr of town (North Sea) gas will cover most normal brazing jobs and the next size up, three times as large, fits the same handle and will tackle almost anything. With a self-blown torch, however, you will need several nozzles. Try to see them in use before deciding, and I strongly advise you to try handling the torch if you can, before purchase – even if it has to be handled cold.

The main problem I have found is that some brands of hose are rather stiff and can interfere with easy manipulation of the torch. There is, of course, no reason why (for low pressure torches anyway) this should not be changed. One final note on this point. Manufacturers are always 'improving' their products and it is an unfortunate fact that this often means that nozzles from last year's model will not fit the current production of handles! Check this when ordering new nozzles for an existing outfit.

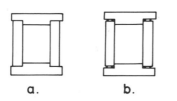

Fig. 65 *(a) Clamping will eliminate the joint gap on machined surfaces. (b) A few burrs or centre-pop marks will establish a capillary gap.*

Holding and Clamping I shall be talking about joint design in the next chapter, and one of the basic rules is that where possible the design should be such that the parts 'stay put' whilst being brazed. However, this is not always possible, and it is often necessary to hold parts together by external means. The important thing here is that any device you use maintains the integrity of the joint gap. In a joint such as Fig. 65A, if the surfaces are machined and a clamp applied the joint gap will be zero and you will get no penetration. It is no use filing little channels for the alloy to get through to the other side, for even if it does you will have the joint dependent solely on the fillets and these may be small. The trick here is to put a few small burrs on one surface – the upright ones in this case. A few centre-pops will do (see Fig. 65B). One brazier I saw demonstrating simply tapped the edge of the plate on a sharp corner of the vice, thus raising a few burrs which kept the two pieces apart! With brass parts you could tap one part against the other.

In the old days 'Binding with black iron wire' was commonplace – and not without reason. This method, provided you can get the binding on in a convenient place, has the great advantage that it is unlikely to cause any distortion. In Fig. 66 the clamp body will remain relatively cool, but its screw, and the work, will expand. This will bend the work and you will find this distortion remains after it has cooled. Iron wire bound round the centre would get as hot as the work and exert far less force. If clamps *must* be used they should be attached at a judicious spot (see Fig. 67). Two are used, one at each end; had one been applied in the centre of the beams it would have bent the base. (I

Fig. 66 *If the clamp is applied in this fashion the work will almost certainly suffer distortion.*

Fig. 67 *Preferred method of clamping. Distortion is unlikely, but means of assuring the capillary gap may be needed.*

know, because that is what happened the first time I made it!) The clamps I use are small, and in sets of three, ranging from half-inch up to 1½ inch in the gap. All steel, they have stood p to repeated heating – and to being immersed in pickle – for a good many years now.

Fig. 68 shows the small 'third hand' I use for brazing. This is rather more robust in the grip than that shown in Fig. 48, and the fingers are tightened onto the work with small screws. Its construction is described in 'Simple Workshop Devices' and it is well worth making one. Though normally used with a fixed torch it can be set in a small vice and used in the normal way. It can be regarded as a 'universal jig' on a small scale.

Tailpiece With any form of torch you will have to 'light the gas'. The flint-and-steel device is now universal – indeed, many outfits include one as part of the

Fig. 68 *A 'third hand', rather more robust than that shown in Fig. 48.*

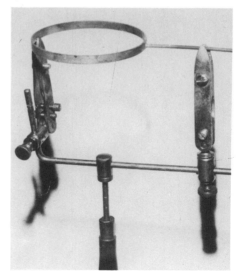

kit. Or there are piezo-electric ones and even some with 'micro-chip control'. I have one or two of the flint type, and sometimes use them, but find that all are unreliable and some quite hazardous; if they don't work first time you discharge quite a body of gas trying to get it to work and then – 'whoooomf!' At my age I value my hair rather more than most people! So, I use a little paraffin lamp – actually one of the little night-lights that used to be common, with a wick about ¼ inch diameter. It lasts for hours and never fails to ignite the gas. Better than a candle (which I have also used) as it doesn't spill wax all over. The only points to watch are, first, that the lamp is sited so that, when the torch does light, the flame doesn't impinge on anything that matters; and, second, that it is not situated where you might inadvertently direct the torch whilst brazing. In passing, don't use a *gas* pilot light in this fashion. Occasionally the torch may blow out the 'lighter' flame, and it is all too easy to forget to turn it off. With a paraffin wick there is no danger.

Chapter 10

Capillary Joint Design

The design of a soldered or brazed joint is just as important as the selection of the alloy and flux and its application – sometimes even more so. Naturally there are many occasions when the *shape* of the finished fabrication is dictated by the overall design of the finished article but even here it is prudent to give thought to the detail *joint* design as early as you can. In most cases it is possible to achieve a reasonable design or, at worst, avoid the most serious faults. As usual, a sense of proportion is desirable; a simple water-pot or a dust cover need not exercise our attention as much as a pressure vessel or a load-bearing beam. I cannot design your joints for you, as so much depends on they type of model you are making; nor can I cover the field by making case-studies of the many types of joints. However, I hope that by outlining the principles involved, and then describing the more common situations, I shall at least give you something to go on.

Principles The main principles to be observed are:

(a) The integrity of the joint-gap must be maintained between the upper and lower limits proper to the alloy being used. Alternatively, if the correct joint gap cannot be assured, then an alloy must be selected which can accommodate the variation.

(b) Where possible the joint should be such that the mating parts will 'stay put' during the soldering or brazing operation without the use of clamps. If clamps or jigs MUST be used then care must be taken to ensure that the clamping pressure will not cause distortion, or that the mass of material in the jig will not rob the joint of heat.

(c) If there is likely to be any differential expansion between the parts being joined the effect of this, particularly on the joint gap, must be considered. In general when two parts fit one inside the other it is best to arrange for the alloy to be compressed as the parts cool, rather than the reverse.

(d) Loads which may fall on the joint should preferably impose a *shear* stress, though (with brazed joints) tensile stresses can usually be accepted. 'Peel' or 'tearing' loads should be avoided, and any bending effects should be examined to ensure that no peeling stress will be imposed.

(e) Where possible the stresses should be carried by the alloy within the capillary gap. Fillets may reduce 'stress raisers' or inhibit a peeling action, but should not be relied upon to carry loads by themselves.

(f) The joint should have no 'dead end' in which flux can be trapped – there should always be a clearway through the gap so that the alloy can drive out the flux before it.

(g) Provision must be made for the escape of air or hot gases from any closed container; if need be an escape hole must be provided which can be sealed up later.

These principles apply equally to soft solder and to brazing, but the former is far less sensitive to excessive joint gaps. Soft solder is, however, much more susceptible to peeling action, and this must always be kept in mind. Fig. 69 illustrates this point. The force acting on the stem of the 'Tee' at 'a' will cause peel at the junction, and this is aggravated by the rounded corner of the flange. The disposition at 'B' will be better, and even more so if the joint is fed from the bend side so that a fillet can form, as at 'c'. We shall deal with tee-joints in more detail later. It is, of course, axiomatic that soft solder will

be weaker than brazing in any joint; brazing alloy is as strong in the joint as most base (parent) metals, but even the beat of lead-tin alloys will not return better than 4 tons/sq.in. Further the 1000-hour creep strength of normal tin-lead solders at 100°C is very low indeed.

Butt Joints The plain butt-joint (see Fig. 70) is seldom used. Not necessarily because of any weakness, but rather

Fig. 70 *Butt joint.*

because it is very difficult indeed to maintain the joint gap during the making of the joint. When brazed, a joint such as this between two rods will, if alignment has been maintained, be as strong as the parent metal (though the scarf joint, Fig. 77, is better), but in sheet work it is used only on decorative ware — copper vases and the like — where the skill of the coppersmith is called into full play! The *Stiffened Butt Joint* is, however, very common (see Fig. 71). Here a butt-strap is fitted, giving a capillary joint which will act in

Fig. 71 *A butt-strap strengthens the joint. W = 6 to 8t for brazed joints, up 10t for soft solder.*

shear, greatly strengthening the joint. Provided the joint gap is maintained it can be located temporarily by screws or rivets, filed off later, and so presenting a secure assembly whilst soldering and brazing. Unlike a riveted lap joint there will be no tendency for the joint to open up under stress, *provided* that the gaps

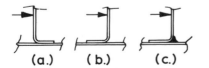

Fig. 69 *Peeling. (a) The load will tend to peel the joint apart. (b) Peeling action is almost eliminated if the lap is reversed. (c) A substantial fillet improves matters still further.*

are properly filled with alloy. Most butt-strap designs published present some doubt about this, because the width 'w' is too great. The usual recommendation here is that the lap (=½w) should not exceed three or four times the thickness of the plates to be joined. The strap itself should be of the same thickness of material as the plates. Fig. 72 shows the *lock-joint* universally found on the sides

FLUSH SIDE.

SEAM SIDE.

Fig. 72 *The Lock joint. Once closed up the parts are automatically retained in place.*

of tin cans. This is especially useful for solft-soldered joints, as filling of the gap is virtually certain, but it can be used for seams in small brazed pressure-vessels of thin material. The joint is usually made flush on one side. If made for brazing the joint is first hammered up and then 'jarred' at one end to loosen it a trifle and a free-running aloy is applied. The rod is applied as shown by the arrow and the joint then inspected

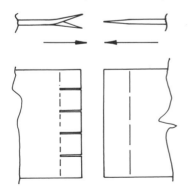

Fig. 73 *The Brazier's joint. Note the thinning of the sheet edges.*

at the opposite side to see that full penetration has been achieved. I have used this joint very successfully on small model boilers rolled up from 24 gauge material, both copper and brass.

The Brazier's Joint (see Fig. 73) is very strong indeed, being, in effect, a series of fined-down lap joints on alternate sides of the sheet. The edges are first

Fig. 74 *(a) Brazier's joint prepared. Note that some 'tabs' have been cut trapezoidal in shape. (b) The joint assembled and hammered close. (c) After brazing. The trapezoidal tongues give the impression of dovetails.*

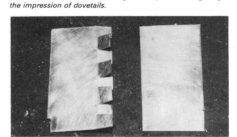

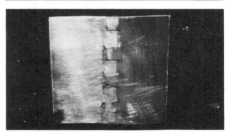

Fig. 75 *(a) The outside of a brazier's joint which has been hammered and filed flush. (b) The inside of the same joint.*

thinned down and then a series of slots cut in one of the mating parts. The flaps so formed are bent alternately upwards and downwards, the two halves brought together and the flaps then hammered down. If desired the flaps on one side can be hammered sufficiently to embed them slightly into the mating

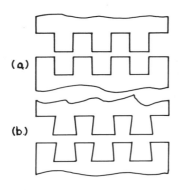

Fig. 76 *Two forms of 'Dovetail' joint.*

sheet. The joint is then jarred to loosen it slightly and so ensure that there is a capillary gap throughout. After brazing one side may be filed flush. Fig. 74 shows the stages in making such a joint (this shows some of the laps in the form of 'dovetails', favoured by some, but offering little except visual advantage) and Fig. 75 the inside and outside of a joint which has been filed flush. The virtue of this joint springs from the fact that even in a large assembly the brazing alloy has a relatively short distance to travel to fill the capillary completely. It can be equally useful with soft solder.

The Dovetail Joint (see Fig. 76) has a similar appearance, but in this case the two plates are cut to matching and interlocking shapes. These may be plain as at 'a' or true dovetails as at 'b'. Many practitioners use plain 'tails' over most of the length, with a proper dovetail at each end, these latter helping to hold the plates in the correct position during the brazing (or, of course, soft soldering) without the need for clamps etc.

This type of joint is very strong, but presents considerable problems in the preparation. The cutting must be done very carefully and it is really a 'fitting operation' to ensure that a reasonably uniform joint gap of the right size is maintained. It is, perhaps, a relic of the 'spelter' brazier's art, when joint-gaps were not quite so important. On 'production' work in industry, of course, the cutting of the matching pairs could be jigged and done with stamp-shears, obviating most of the difficulty.

If the material is so thick that the overlapping type of brazier's joint would be impossible then the **Scarf Joint** (see Fig. 77) can be used and is just as effective. The scarf can be machined – and this is preferable – and a few

Fig. 77 *The Scarf joint.*

screws used to hold the two parts in line. Centre-pops will preserve the joint gap. The length of the scarf should be between three and four times the plate thickness, though for joining two shafts or rods together end-to-end an angle of 60° would probably suffice.

The Cranked Lap Joint (see Fig. 78) is useful for its simplicity. The main problem here is assuring dimensional accuracy (as in the case of the lock joint) but where this is not of paramount importance it has the merit that one side of the joint will be flush with no extra

Fig. 78 *The Cranked Lap joint. The lap may be wider if temporary screws or rivets are used.*

work. The width of the lap should not be overdone – 6 to 8 times the sheet thickness for soft solder, and 3 times the thickness for brazing, though in the latter case I do make it a little wider if rivets or screws are to be used to hold things together whilst heating.

The Tee Joint This is the most problematical of all capillary joints, whether soft-soldered or brazed. The direct butt (see Fig. 79) is, unfortunately, some-

Fig. 79 *Plain tee-joint. Fillets are desirable when brazed, essential with soft solder.*

Fig. 80 *'L' joint, with a lap formed from the base sheet. Compare with Fig. 69.*

times unavoidable. It is inherently weak being subject to shear, tension and peel at the same time if there is any sideways load and this is bound to be present if the joint forms part of a boiler or similar pressure vessel. If the design cannot be adjusted, then every effort must be made to form a really good fillet, and this may well demand that the order of brazing be altred so that a higher melting-range alloy can be used at this point. Fortunately the situation usually arises in joints where the plate is fairly thick (⅛ inch upwards) when the difficulty is not so serious. However, with thin material even a small flange, as at Fig. 69B, will help materially. For brazing the flange need be no wider than three times the sheet thickness. Occasionally the flange may be set on the base metal, as in the 'L-joint' in Fig. 80.

In most of these cases there is a second problem – that of holding the parts together. I use the method shown in Fig. 81, which has the secondary merit of strengthening the joint considerably. Tenons are machined on the

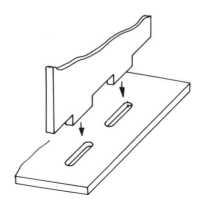

Fig. 81 *Method used by the author both to strengthen and to assure accurate location in Tee-joints.*

'leg' of the tee and slots in the base. The joint-gap is assured by making a few centre-pop marks and these also help to lock the two parts together. On thicker material, of course, the two parts can be held together with small screws.

Boiler Tubes. Opinion on the best method of fitting boiler tubes – or, indeed, any round part which fits a hole – is very divided. The general tendency seems to be to make them a good fit, and then to file a few notches so that any brazing alloy applied on one side will run through and form a fillet on the other. This is not satisfactory, for it means that within the joint itself there may – almost certainly will – be no bond at all; worse, there may even be trapped flux. The fillet size is indeterminate, and although the joints so made may not leak the strength overall arises solely from the fact that there is usually a fair number of tubes. In the case of a boiler the joint cannot be examined internally. The proper way of proceed-ing is to ensure that there is a *designed* joint gap and then to fill it. The ideal method is to ream the holes in the tube-plate and then to machine the tube ends to a designed joint gap.

If this is done the capillary action of the molten alloy will centre the tube in the hole automatically – see Fig. 82 – there is no need to make any provision to centre the tube so long as it is free to move. If for any reason you suspect that it may be forced sideways (e.g. if the tubes have already been brazed into the other tube-plate) then a few centre-pop marks around the tube – four will be enough – will look after matters. There are many ways of preventing tubes from falling through the holes – rolling in a ridge, bell-mouthing the ends etc, but the machining method is just as effective. The *really* important point is that there must be a joint gap of at least the minimum radial width which is appropriate to the alloy being used. If you wish to apply a fillet as well this can be done simply by feeding more alloy, and I allow for moderate fillets when calculating the size of wire to be used. (I always use preformed rings for tube brazing.)

In industry use is made of the diamond knurl for this purpose. The points of the diamond pattern locate the part and it does not matter that these may form an interference fit, for there is adequate space for the alloy to flow around them. The size of the pattern must, of course, be proportioned to the thickness of the plate, and the method is limited to work which is strong enough to stand the knurling pressure.

Blind Holes There are many cases where one part (e.g. a shaft) must be brazed or soldered into a socket. If the design is as Fig. 83a there are two

Fig. 82 *Showing the 'self-centring' action of the surface tension within the liquid alloy in the joint-gap. The tube is ⅜ inch diameter.*

unfortunate consequences. First, the alloy cannot displace either the flux or any vapour or air which may be trapped in the capillary. Second, there is no means of assessing whether the alloy has in fact penetrated the joint at all. A hole of reasonable size at the bottom of the socket, as at 83b, may effect a cure, provided that the whole joint can be made hot enough to ensure capillary flow throughout. The best solution is shown at 83c. Here the alloy is placed as a small pellet *within* the base of the socket. No more than light pressure is needed on the spigoted part – often its own weight will be sufficient – but a slight pressure IS needed. This is because the air trapped in the cavity will expand on heating and tend to eject the spigot. Proper penetration of the whole joint is assured when the bright ring of alloy appears at the mouth of the joint.

Fig. 84 shows sectioned photos of such a joint; in one there has, in fact, been reasonable penetration, but there is a cavity part-filled with dead flux below. In the other, when a pellet was set as in Fig. 83c, the whole joint is filled and there is but the barest trace of a pinhole inclusion at one point. This method of applying the alloy – whether soft solder or brazing rod – can be

Fig. 84 *(a) Section of joint made following the design of Fig. 83(a). (b) Section of a joint made with alloy set in the base of the cavity – Fig. 83(c).*

extended to joints of many shapes and configurations with advantage; not the least of its merits is the fact that the alloy is never exposed directly to the flame.

Tube Closures Fig. 85 shows another application of the method just described. The usual type of closure for the end of a tube using hand-fed rod is shown at 'a'. Fortunately the lap is usually short and we can be reasonably

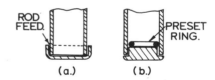

Fig. 85 *Two methods of closing a tube end. (b) is preferred.*

sure of good penetration, but inspection is impossible if the tube is of any length. At 'b', however, with a preform in the shape of a ring set inside the joint we not only have visual evidence of the penetration, but also make a neater joint and save solder or brazing rod. It is an established fact that joints of type 85a (and those of Fig. 83a or 83b) are

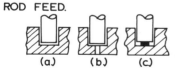

Fig. 83 *(a) The flux cannot escape and will be trapped in the base of the socket. Poor penetration of the alloy is likely. (b) An alternative design, providing a hole through which the flux can escape. (c) The preferred arrangement, with a piece of brazing alloy set in the cavity.*

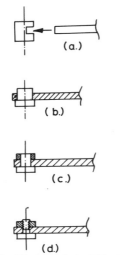

Fig. 86 *Four methods of fitting a boss to a link or lever, for brazed or soldered construction.*

almost universally overfilled, resulting in a large and unnecessary fillet, when hand-fed with rod.

Levers and Links The fitting of a boss at one or both ends of a link is one of the commonest applications of 'small brazing' or even soldering processes. Soft solder is rather neglected in this instance, but if the object of the boss is not more than to provide an extended bearing surface it will be quite strong enough. However, most cases do demand brazing and in any case brazing is often easier.

Fig. 86 shows three common and one not so common constructions. At 'a' a slot is milled across the boss, the arm being made a slack fit to provide the necessary gap. It needs but little care in machining both to ensure that centre-distances are held – at least within a few thou. However, a slack fit can lead to difficulties in locating the parts during brazing, and my own practice is to make the slot fairly deep and slightly narrower than the arm, the end of the latter being filed to a slow taper; when the parts are tapped together the taper wedges the parts sufficiently to hold them, but still provides a capillary gap. At 'b' we have the simplest method, which has the slight disadvantage that the boss is of different diameter on each side. This difference need not be large. When bosses are needed at both ends of the link the centre distance can be held exactly if the holes are drilled or bored using co-ordinate methods. 'c' is a variation, ensuring that the bosses are equal in diameter. If the hole in the collar is made undersize and then treated with a taper reamer it can be wedged during assembly. 'd' is a method I have found useful, as it needs but a small hole in the lever. The thread is usually made smaller than the hole eventually drilled through the boss, so that it disappears. I make this thread a slack fit, and set a few centre-pop marks on both sides of the link to provide the capillary gap and the 'nut' can then

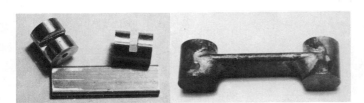

Fig. 87 *(a) Example of the design shown in Fig. 86(a). (b) The link completed. The joints have been over-filled to provide large fillets. The bosses are at 1.2 inch centres.*

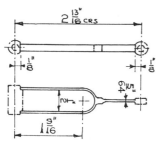

Fig. 88 *A connecting rod for a model of an early type of steam engine.*

be tightened down and hold all secure whilst brazing. In all cases it helps to set a small centre-hole at one end of the boss before assembly, to facilitate the final drilling. Fig. 87 shows a small lifting ink made using method 'a', the joints having been over-filled to give a fairly large fillet.

Fig. 88 shows a more difficult example – the forked connecting rod for a model of an early 'overcrank' engine. The main body is a forging, the construction of which need not concern us, though it was 'interesting'! The bosses were brazed on afterwards and correct alignment of the two lower ones was important. The ends of the fork were, therefore, carefully machined dead square across, and the two bottom bosses made as one, with slots milled as in 86a; the top boss also carried a slot. The whole was then assembled as in Fig. 89. The alloy used in this case was Johnson Matthey's 'G6', which is specially formulated to give a silvery appearance, 67% silver and with a fairly short, but high, melting range. Surplus metal between the two arms was then machined away.

Fig. 90 shows the beam for a small beam pumping engine model, only 5 inches long. The bosses were fitted using method 86c and brazed. The

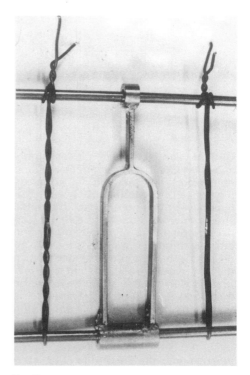

Fig. 89 *How the parts for Fig. 88 were assembled for brazing. The surplus of the lower boss was cut away later.*

flange round the 'circumference' of the beam was then soft-soldered in place using BS.219 grade 'C' alloy, melting range 183-227°C. The ribs, made from half-round brass strip, were tinned, as was the body of the beam, and then soldered in place using 'electronic' quality cored solder, BS.219 grade 'A', melting range 183/185°C. Both soldering jobs were done with the iron illustrated in Fig. 42. This method can be used to make metal patterns for foundry service. This beam is seen again in Fig. 92.

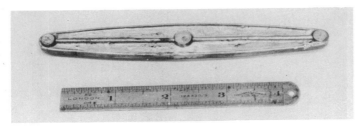

Fig. 90 *Both brazing and soft soldering were used in the construction of this beam.*

Girders and Beams The assembly shown at Fig. 91A is very unstable, and not to be recommended. 'b' is better, the edges of the top and bottom members having been machined; the detail shows how the edge of the flange can be tapered off to give the impression that it is thinner. If the box is to have no projecting flanges the arrangement on the left would be used, but I would prefer to leave just a little overlap, to be machined or filed off afterwards. 'c' is better still, and, again, the projecting flanges can be thinned down if desired, or machined off if a flush finish is needed. In the latter case the groove can be very close to the edge – its purpose is only to locate the verticals. In making beams of joint section, as at 91D, stability is difficult, but a few screws can be used to hold all together. A few centre-pops may be used on the edges of the web to provide a gap but often enough the slight burrr left when tapping the small holes will suffice. In all such designs the raw stock should be stress relieved before machining, as with bright drawn steel or extruded brass the locked-up stresses will almost certainly cause some distortion during the brazing. Bearing in mind that the stress on the joints of a beam made like this are entirely in shear the use of a good quality soft solder – perhaps the silver/tin/lead alloy – might be considered instead of brazing, with considerably less risk of distortion.

Fig. 92 shows a 'mock-up' assembly of a model engine made in prototype by fabrication and later translated in castings ('Lady Stephanie'). The top entablature is a fairly complex beam structure, and Fig. 93 shows the procedure. The basis is the top flange, which is cut

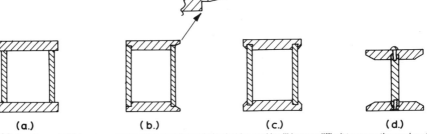

(a.)   (b.)   (c.)   (d.)

Fig. 91 *Box Beams. (a) This arrangement is liable to collapse during brazing, and it will be very difficult to assure the spacing of the webs. (b) A better arrangement. 'Flush flange' at the left, flange thinned down at right. (c) The best arrangement; assembly will be sturdy and location is secured. (d) A combination of machined grooves and 'assembly screws' is often necessary.*

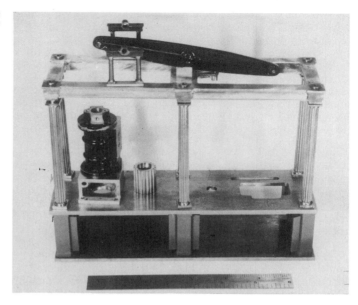

Fig. 92 *The entablature of this model was fabricated, though later produced as a casting.*

from a stress-relieved sheet of brass, though steel could have been used. Locating holes were then drilled at each of the column centres. The six square capitals 'b' are machined with a projecting boss which fits the holes in the plate

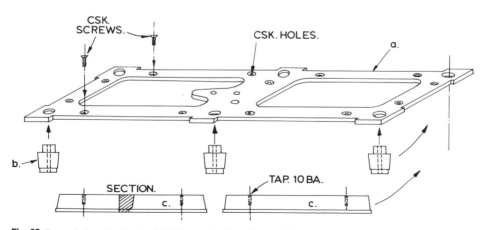

CSK. SCREWS.

CSK. HOLES.

a.

b.

SECTION.

TAP. 10 BA.

c.

c.

Fig. 93 *The methods used to locate and hold the parts for the entablature of Fig. 92.*

'a'. There is a drilled pilot hole through each boss. The side and crossbeams, 'c', are profiled and the ends shaped to be a snug fit between the capitals, and held to the plate 'a' with small screws. Once assembled the beams 'c' lock the capitals 'b' in place, these also being located by the bosses. The boss and rib on the governor shaft support at 'e' are also located by spigots and small countersunk screws. The whole was then brazed up with AG2 alloy at one heat. After pickling and correcting a slight twist, certain small pads (not seen in the photo) were attached with soft solder, and the holes in parts 'b' opened out to receive the spigots at the ends of the columns. The profile of the beam sections is quite ornate, but presented no problems as these were machined in a length and then cut up to size for the fabrication. The use of a single plate as the top flange of the beam assembly greatly eases fabrication and the only debatable point is whether it might have been better to locate the capitals using screws instead of relying on the captive effect of the beams and on the bosses for location. Incidentally, the screws were very small – 10 and 12BA – as they had no work to do other than hold bits in place.

Fig. 94 is an example where machined grooves have been used extensively. It is the crosshead-guide for a rather unusual model engine having a very short stroke relative to the bore. The guide 'a' is a piece of cast GM, and is machined on the outside to dimension, with a short spigot at each end, and a hole bored through to a size which provided a machining allowance for the final bore. The end-plates 'bb' are machine bored to give 0.003 inch clearance on the spigots, providing a 1½ thou radial gap. The base 'c' is grooved to receive both the centre web, 'd', and the two endplates, and a longitudinal groove is cut in the guide cylinder 'a' as well. Part 'c' is over-thick to provide a machining allowance on the face. In fact, the fabrication is treated exactly as would be a casting once it is brazed up.

Though all parts are 'located' the assembly is not rigid, as all the slots are made to provide a capillary gap. This was not strictly necessary, as fillet-

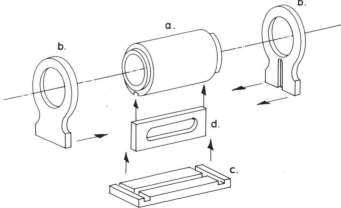

Fig. 94 *Considerable use has been made of machined grooves in assembling this crosshead guide.*

brazing would have been quite strong enough, but the clearance was there, so that some clamping was needed. The obvious place would be across part 'a' and the basic 'c', but thermal expansion would almost certainly have caused the base to bend upwards a trifle and, perhaps, even distort the aperture in the web. Two small clamps were, therefore, applied, one at each end, from the bore of the guide to the baseplate. Fig. 95 shows the completed work after machining.

Use of Brazing Paint Fig. 96 shows an eccentric rod fabricated from ⅟₁₆ inch square mild steel rod. The side members carry screwed ends which were turned from the solid rod. The forked

Fig. 95 *The fabrication of Fig. 94 after machining.*

end is machined with two slots ⅟₁₆ inch wide to accept these members and this joint was brazed first, using AG1 alloy. The members were then bent to the desired shape and the screwed ends secured at the correct distance apart with a strip of scrap steel in which two holes had been drilled. The two diagonal stiffeners were then bent to fit between the side members and one of them filed to make 'halved' joints where they crossed. These joints were also brazed up with AG1. The whole was then assembled and bound with black iron wire (black, to reduce the risk of it 'taking the braze') to maintain a slight tension on the inner stiffener pieces, and the eight joints each anointed with a blob of brazing paste. The flame from a pin-point burner (Fig. 63A) was then

applied to the rods adjacent to each joint in turn until that particular blob was seen to run. The dummy spacer at the screwed ends was, of course, kept in place during the brazing. No difficulties were met during the actual brazing, but the final cleaning of the rather complicated shape was somewhat time-consuming.

As a final case-study, let us look at the example of step-brazing already referred to – the bottom section of the boiler shown in Fig. 23, page 35, and in Fig. 51, page 76. As already explained, this has to be fabricated in three main sections so that some intermediate machining operations could be carried

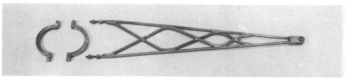

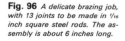

Fig. 96 *A delicate brazing job, with 13 joints to be made in ⅟₁₆ inch square steel rods. The assembly is about 6 inches long.*

Fig. 97 *A 'developed' copper sheet, and the former over which it will be shaped.*

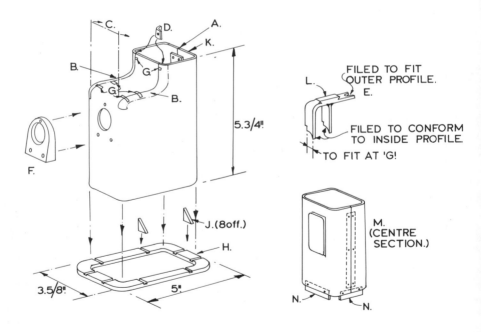

FILED TO FIT OUTER PROFILE.

FILED TO CONFORM TO INSIDE PROFILE.

TO FIT AT 'G!

M. (CENTRE SECTION.)

5.3/4"

J. (8off.)

3.5/8"   5"

Fig. 98 *Illustrating the sequence of brazing on parts of a copper boiler.*

out; the complete boiler was too large to be accommodated on the machines. Fig. 98 shows the details.

The first step was, of course, to 'develop' the sheet metal (copper) in the flat, and also to make a wooden former over which the flat sheet could be beaten to shape. Both are seen in Fig. 97 – the calculator was needed to work out the allowances for the various bends! The next step was to bend the metal round the former and 'fit' the edges of the vertical joint ('A' in Fig. 98) so that there was a uniform gap full length. Holes for the butt-strap rivets had already been drilled, and these were used to nail the edges to the former whilst the side radii at 'BB' were beaten over. (The leaf 'C' was in the position shown chain-dotted.) Three intermediate anneals were needed, but fortunately the little nails still held after the last one. The edges of the side-plates at 'BB' were then filed so that the leaf 'C' would be a close fit and this was folded down for trial.

After fitting the butt-strap at 'A' this joint was brazed using a 16% silver alloy, range 790-830°C (Johnson Matthey 'Silverflo 16). After pickling the shell was carefully corrected for distortion and the front leaf 'C' properly fitted. Small butt-straps 'DD' were fitted at the top of the junction at each side. These two seams were then brazed with DIN L-Ag25 (Johnson Matthey Silverflo 25) melting range 700-800°C. This range slightly overlaps that used for the previous step, but as the joints were remote from the rear seam 'A' there was no reflow there.

The two bearing pedestals shown in the detail were made, though this is really diagrammatic as the two pedestals not only differed in width but also had differently shaped feet. These were

Fig. 99 *Showing how the bearing pedestals are located in slots machined in the casing. The firehole door will be secured with shorter screws before brazing.*

Fig. 100 *The casing ready for the third stage in brazing. Note the heavy cross-bar. The clamps are all-steel, and have survived scores of brazing heats.*

Fig. 101 *The base flange and ribs.*

brazed up to AG5, as good fillets were needed to accord with the shape of those on the prototype, The firehole door 'F' was also prepared together with a matching piece which fits inside the shell, and the holes drilled for the temporary fixing screws. The top of the shell was then machined with slots at 'GG..GG' to receive the spigots of the pedestals (see Fig. 99). Fig. 100 shows the fire-hole door and pedestals in place, fluxed ready for brazing. The very heavy strap clamped across the pedestals serves both to keep the faces in line and also to act as a 'heat shunt', to reduce the risk of remelting the brazing alloy used there. AG14 was used to braze both pedestals and firehole door-frame at one heat, and the job was then pickled again.

The baseplate, 'H' in Fig. 98 (and shown also in Fig. 101) is machined with slots to accept the eight ribs. These were made a drive fit to the slots, as a fillet-braze here is quite adequate. After squaring up, both flange and ribs were brazed to the shell with AG1 (see Fig. 102). The shell was a fairly tight fit in the hole in the flange, and 'stayed put' during heating with no difficulty. No clamps were used.

After pickling again the machining was carried out — facing and drilling the top surface of the bearing pedestals, 'L', and trueing up the edge 'KK'. The centre section of the body, 'M', had previously been machined on the corresponding edges, and fitted with part-length butt-straps, 'NN', to engage within the rim at 'A'. The top section of the boiler involved a similar sequence of operations and rather more intermediate machining (the engine cylinder is carried within the top section of the boiler) and this was finally machined on its lower edges and fitted with short butt-straps, to mate with the top of the centre section, part 'M'. The three parts were finally brazed together, the top section first, using AG2, with a through-bolt holding all together during the heats.

Altogether thirteen separate heats were involved on the whole boiler, and no difficulty was found in correcting the few slight distortions which occurred; there was a slight tendency for the lower and centre sections to take up a trapezoidal shape. Hindsight suggests that two of the brazing operations could have been carried out at the same time as the previous set, but if this *had* been done and any distortion resulted it would have been very difficult to correct. Had it been possible to machine the parts after assembly the number of brazing operations could have been reduced to nine. Fig. 103 shows another view of the finished fabrication.

The 'design lessons' from this example are fairly clear. The first joint was 'located' with butt-strap and rivets. The second joints to be brazed were remote from the first, and the ends were located in the same way. The third step involved three parts, but all were positively located either by screws or by interlocking, and a clamping bar was used on the pedestals to maintain align-

Fig. 102 *Brazing the base flange and ribs.*

ment. However, as they were, perforce, set literally on the previous joint an alloy with a definite melting range gap from the previous one was used. Further, to protect the brazing on the pedestals a heat shunt was applied. The fourth joint was remote from the third one, and the fifth even further away from the fourth.

Conclusion To cover the whole gamut of capillary joining by relating case-histories would need a book in itself – probably more than one volume at that! But I hope that those which I have shown will give some idea of general procedures, procedures which can be adapted to many other types of fabrication. I would commend to you the use of 'preforms'. (A number were applied in making the top section of Fig. 101.) In all cases where the joint can be heated indirectly and a reasonably short melting-range alloy is used, preforms give a far better placement of the alloy than any hand-held rod. Further, where many joints are to be made – as for the tubes of a large boiler – a considerable

Fig. 103 *Another view of the boiler – see also Fig. 23.*

115

economy on brazing rod can be effected.

I would also commend to you the practice of actually *machining* mating edges of joints where this can be done. This helps to ensure that the capillary gap is correct and, perhaps more important, greatly assists in assuring the dimensional accuracy of the work after brazing. In the same connection, though I have not mentioned this matter as such, I would remind you to allow for machining when designing the fabrication! With the best will in the world you must expect some change of shape after heating a complex shape to brazing temperature. If you design the piece so that it is subsequently treated as you would a casting you will not go far wrong.

Finally, you may feel that I have used the words 'capillary' and 'gap' far too often. 'Often' – true, but 'too often', by no means. Without such a gap you can neither braze nor solder properly. It may seem to be an excess of zeal to apply a micrometer to parts which are to be heated red-hot and filled with molten brazing alloy or solder, but it does pay dividends and you will find far fewer pinholes or leaky joints if you get the gap right.

Chapter 11

Safety Precautions

Almost everything we do carries some hazard with it, but we survive because we take 'reasonable care'. If we tried to erect guards against 'every known risk' life would be one long imprisonment. But if we did NOT take care the story would be very different – very little work would ever be finished. So, the first rule must always be to identify any hazards, assess the risk factor, and then take steps to reduce these risks to acceptable proportions.

In both soldering and brazing the main areas of risk are three. *First*, heat. Even soft solder will cause a nasty burn and a soldering iron can cause a fire. *Second*, from the materials used. Most fluxes are irritant, if not poisonous. *Third*, from fumes from the heated and decomposing fluxes and, in some cases from one or two of the alloy constituents. Let us look at these in order.

The risks from 'heat' lie in two directions. You may set the place on fire, and you may burn yourself. When using soldering irons there is always a chance that the hot bit may contact some flammable substance, and this must be guarded against. One of the advantages of the Superscope iron (see Fig. 42) is that it turns itself off as soon as you put

it down. But all irons (including that one) should have a rack or rest on which it can stand, and you should make it a true habit always to use it. The bench top should be covered with something that is not only heat resistant, but which will also prevent heat from passing through to the bench.

Brazing torches are far more dangerous. On a large one the flame itself may be a couple of feet long, and the region of hot gas a further foot or so again. It is all too easy during the brazing operation to move the flame aside for a moment and fail to notice that it is pointing in a dangerous direction. The first and simple precaution is to arrange things so that such a thoughtless action can do little harm – don't braze inside a wooden shed! Second, get into a rhythm so that *as* the torch is moved aside you glance at it automatically to see where the flame is going to.

A very serious risk indeed can arise if a torch flame finds a gas hose. True, there is (or ought to be) a hose-failure valve incorporated in the regulator, but a 9-foot hose pressed at 56 lb/sq.in contains a considerable amount of propane and the rush of flame can be disastrous. Finally, lighting the gas – if the torch does not ignite straight away you can find a considerable volume of fuel ready to blow up as soon as ignition *is* achieved. In my opinion all large torches should have both a pilot flame *and* a trigger action, so that the main nozzle is not activated until the pilot has lit. I have already referred to the risks attached to spark igniters and the like, but repeat what I said before; these are fine for small flames, but for anything larger than a 4 inch I prefer to use a permanently lit paraffin flame to light up from.

Now, if you are going to set light to the shop, what to do? First, have means of first aid — even a bucket of water. Don't forget that you have hot metal about the place as well as torch flames; a CO_2 extinguisher will possibly put out a fire in burning material, but it won't stop a red-hot boiler firebox from burning its way through the floor. A judicious application of water will. But don't rely on D.I.Y. methods if you have a serious fire — dial 999 for the professional. Opinions differ on what to do if a gas hose flames up and the failure device doesn't work, or if a torch gets out of control. The worst thing to do is to extinguish the flame, for if raw gas continues to escape an explosion is almost certain. The best plan in my view is to use your common sense and whatever means is available to get at the main gas valve and turn it off. You can put the fire out afterwards.

However, it may be that the workshop is safe, but you have burnt *yourself*. The first thing to remember is that the burn itself is not the main risk — it is the filth on your hands which provides the hazard. Dirt round the burn will infect it, but a clean burn can soon be treated. Cold water will do no harm at all to a burn — rather the reverse. So, apply lashings of cold water and get someone to wash the area around the burn, carefully. DON'T apply any of the sticky ointments you may have about the house; a *dry* burn dressing will do little harm, or a pad made from a clean handkerchief or pillowcase (never surgical lint or cotton wool) held in place with a loose bandage. The object is simple — to keep all dirt off the burn. Then if it is other than a trivial one hop along to 'casualty' at the local hospital. Leave the treatment to the experts. Don't neglect burns; remember that despite all the clean air acts the atmosphere is loaded with bugs that will infect it in no time at all. (That is why you need the protective pad). Finally, if it is a really bad one, bear in mind that you are then in what is technically known as a 'state of shock', so keep warm and no matter how well you feel DON'T try to drive your car; send for an ambulance or get someone to take you to the doctor or hospital.

However, a sense of proportion is needed! If all you have suffered is a small burn, apply one of the acriflavine based burn dressings (of which you should have a stock in the workshop) and see how it is in the morning. But, I repeat, it is not the *burn* you have to worry about — in the old days they used to cauterise wounds to *prevent* infection — so much as the dirt around it. So, keep it clean.

Hazards from Materials All fluxes should be handled with care. The acid-based soft-solder flux is poisonous, dangerous to the eyes (so avoid splashes) and the fumes can afflict people who have chest troubles. The remedy is simple: keep them in secure bottles which cannot tip over and avoid like the plague those which have 'patent caps' which come off with a jerk; wash your hands after using them – try to avoid touching fluxy surfaces if you are skin-sensitive – make sure they are labelled and keep them out of reach of children. Brazing fluxes can be very irritating, and one of the minor reasons why I like the paste fluxes is that there is no need to handle irritant powder. There is also less risk of the fine powder flying up in the air and getting in your eyes. Almost all these fluxes are ranked as 'extremely irritating' to the eyes and the recommendation if this should happen is to irrigate the eyes with water for at least ten minutes. (Have you got an eye-bath in your cupboard?) All are moderately poisonous if swallowed, and they should be kept away from the larder and from children; the powder does look exactly like icing sugar. If any is swallowed the first aid is to drink lots of milk in which powdered chalk has been mixed, but call the doctor straight away. Do NOT try to make the patient vomit.

Flux Fumes Overheated brazing flux can cause fumes which are irritating, but in normal use there is little risk. Prolonged exposure can cause discomfort but few of us are likely to stand over the job for hours on end. If such discomfort is felt – perhaps due to your being more sensitive than usual – then goggles will help and you will also find the lanolin type barrier creams useful. But the short answer to all fume problems is ventilation – provided that the draught blows the fumes away from you and not into your face. In my experience the worst *discomfort* comes from the vapour of hot 'spirits of salts' soft soldering flux and as the basis is hydrochloric acid a sharp dose could be damaging. For this reason I always avoid the old-fashioned practice of dipping a hot iron into 'killed spirit' fluxes.

Overheated Alloy There is little risk from soft solder, though lead IS a poison and severely overheated soft solder could constitute a hazard. A greater risk arises from the CADMIUM and ZINC in brazing alloys. The working temperature when brazing is only a hundred degrees or so below the boiling point of these metals, and cadmium especially can be lethal. This matter came to prominence in 1980 following the sudden death of a user whilst carrying out a brazing operation. True, the operator was using oxy-acetylene, and in far from ideal conditions, but considerable apprehension was generated. Following comment in the *Model Engineer* an article was prepared for that magazine looking at the problems and their solution in some detail. Through the courtesy of the Editor I reprint this article in Appendix III, Page 130. For the user whose consumption of alloy is but a few sticks per year, however, the hazard is slight provided proper precautions are taken – and these 'precautions' are, really, no more than good brazing practice.

First, when using cadmium-bearing alloys the work should be heated to a temperature no greater than that necessary to flow the alloy into the joint. Second, the work should be done in conditions of reasonable ventilation having regard to the rate of usage of the alloy. Melting one stick an hour, then the normal ventilation induced by the torch flame should be adequate if doors and window are open, but at one stick (10 grams) per minute you will be well advised to arrange for positive ventilation or do the work outdoors. The wearing of a simple face-mask in such circumstances is an elementary precaution. Third, when applying the alloy keep your face well back from the work.

These three points emphasise my previous remarks about the inadvisability of using oxygen torches. First, even momentary contact of the flame with the alloy, molten or just red hot, will – WILL, not 'may' – volatilise some cadmium and probably some zinc as well. Second, these torches induce almost negligible ventilation. They 'use' no air, and the total heat output is not sufficient to cause more than rudimentary air movement. Third, the 'natural' attitude of the user is to stand over the torch, whereas with the air- or self-blown torch you will be compelled to stand back; the work is too hot for comfort otherwise.

There is, of course, the alternative of avoiding the use of cadmium-bearing alloys and all manufacturers can offer cadmium-free brazing materials. The range of alloys I use for sequence brazing are, all except the final one, of the cadmium-free type, for the very simple reason that on a fabrication of this sort the first joints must be made at temperatures which actually exceed the boiling-point of cadmium. However, the popular alloys AG1 and AG2 are so very useful, and make the work so very much easier, that it would be a pity to abandon their use when the hazards can be reduced to very small proportions if not avoided altogether by the application of a little common sense and the adoption of 'good practice' in the brazing operation. This is the basis of ALL safety in the workshop, after all.

Pickling The risk here arises from two possibilities. First, that the 'quench', especially with a large workpiece, may splash the pickle fluid into your eyes. The remedy here is obvious – stand well back, and wear goggles. If you do get any in your eye, irrigate copiously with water and call the doctor or go and visit him if nearby. The danger is not only from the acidic fluid, but also from the fact that this pickle solution is absolutely filthy, with all sorts of oxide and sulphate particles as well as mere dirt. The second risk, especially when pickling anything like a boiler with enclosed cavities, is that you may get a scald from boiling fluid. Again, stand well back, be wearing proper overalls (and shoes or boots) and try to manipulate matters so that any squirt will fly away from you. If you do get a scald treat as for a burn above, taking care not to break any blisters which may form.

The final point which I ought to mention is that of contamination by the brazing torch itself. The flame uses up the air in the workshop, and replaces it with a mixture of carbon dioxide, a little carbon monoxide, and a few 'nasty bits of stuff', the latter in very small quantities but very vicious. From the 'Properties of Fuel Gases' given in the appendix you can work out how long it takes to use up the air in the shop! In fact, of course, the ventilation necessary for safety in the previous section should also look after this problem at the same time, but it all adds up. It is just when you start to get a bit dozy from lack of fresh air that you may make the mistake that sets the place on fire or causes a release of cadmium fumes.

Conclusion Reasonable ventilation, reasonable care, the application of common sense, and the cultivation of good brazing practice will reduce both soldering and brazing hazards to negligible proportions. The time when you need to think is when you tackle a job which is much larger, or which may take much longer than usual, that is all.

Appendix I

Properties of some Commercial Brazing Alloys

The following information has been extracted, with permission, from Manufacturers' data sheets, and is presented to assist readers when faced with a choice of unfamiliar alloys by retailers. It should be noted that the fact that an alloy quoted below has no BS 1845 designation may be due to a number of reasons. It may be to some Continental (DIN) or American National specification; to a British Standard now discontinued (the BS designations are those of the 1984 revision) or it may have been formulated to meet some special customer's requirements.

The strength figures, where quoted, are those ascertained on steel specimens in standard test equipment. However, in the case of the phosphorus-bearing alloys, used on copper, no realistic joint test is possible as the copper is automatically annealed during the brazing and usually gives way before the joint. With these alloys, therefore, the strength figures are those for the alloy 'as cast' in ingot or strip form. 'NA'=Data not available at time of writing.

FSB No.	Melting Range °C	% Silver	BS 1845 No.	Tensile Strength ton/sq.in	Elong.[n] %	Joint gap mm	Remarks
Cadmium-bearing							
1	606/648	38	AG3	32	24	.05/.25	Free flowing, a good general purpose alloy, suitable also for some stainless steel.
2	608/620	42	AG2	30	28.5	.05/.15	
3	620/640	50	AG1	30	35	.02/.15	The 'premium' free-flowing alloy. Close joints (0.001"-0.004") essential.
7	595/630	40	AG10	31	NA	.05/.20	Wide melting range, fillet forming, for irregular joint gaps.
14	616/735	24	–	30	15	.08/.25	Cheaper alternative to No. 2, but wider melting range.
16	612/668	34	AG11	32	20	.08/.25	Nickel-bearing, chiefly for brazing carbide tool-tips. Poor flow characteristics.
19	635/655	50	AG9	32	NA	–	
21	605/710	25	–	25	14	.08/.25	Similar application to No. 16, but cheaper and has wider melting range.
Cadmium-free							
17	M.P. 778	72	AG7	22	20	–	Silver-copper eutectic alloy. Not normally used for torch brazing.
4	690/735	60	AG4	25	23	.05/.15	White in colour. Has good electrical properties.
5	700/775	43	AG5	26	10	.05/.15	An alternative to alloy No. 5.
30	625/725	40	–	31.6	24	.05/.20	Low cost alloy for step brazing.
23	700/800	25	AG17	34	13	.08/.25	Ditto, but with lower working temperature.
25	680/770	30	AG16	24	12	.08/.25	Ditto, but shorter plastic range at higher temperature than No. 23.
26	790/830	16	–	28	25	.05/.15	
Cadmium-free (tin-bearing)							
29	630/660	55	AG14	35	24	.05/.15	Excellent cadmium-free alternative to No. 3.
32	650/725	38	–	34.5	14	.05/.20	
33	655/755	30	AG21	NA	NA	.08/.25	
34	640/700	39	–	NA	NA	.05/.15	
Other							
37	625/705	49	AG18	Nickel-Manganese		NA	Specially suitable for brazing cast iron.
40	880/880/895	1.0	–	Silicon-bearing		NA	Low cost alloy of short melting range for use on copper and mild steel.

Copper/Phosphorus (NOT for use on ferrous metals)

	Melting Range °C	%	BS No.			Remarks
6P	645/700*	15	CP1	45	25	Fluid and will fill small gaps, but subject to liquation if care is not taken.
11P	640/740*	5	CP4	39	6	
13P	645/740*	2	CP2	35	5	
10P	705/800	0	CP3	31	2	

*The true liquidus is higher, but the alloys are about 95% liquid at this temperature. Maximum service temperature 200°C. THESE ALLOYS MUST NOT BE USED ON STEEL OR ANY ALLOYS CONTAINING NICKEL. See p. 37.

(B) – JOHNSON MATTHEY METAL JOINING LTD.

JM Name	Melting Range °C	% Silver	BS 1845 No.	Tensile Strength ton/sq.in	Elongation %	Shear Strength ton/sq.in	Joint Gap in mm	Remarks
Easyflo	620/630	50	AG1	28.5	35	10.3	0.05/0.15	Higher ductility, finer fillets, and better corrosion resistance than No. 2 alloy.
Easyflo 2	608/617	42	AG2	28.5	30	10.0	0.05/0.15	Good general purpose alloy suitable for most applications.
Argoflo	608/655	38	AG3	32	20	10.3	0.05/0.25	Fillet-forming, for irregular joint gaps.
Mattibraze	612/668	34	AG11	32	20	11.0	0.075/0.2	(Mattibraze 34).
ArgoSwift	607/685	30	AG12	30	24	10.4	0.075/0.25	For applications where fillets are required, or where joint gaps cannot be closely controlled.
DIN Argoflo	595/630	40	(DIN Ag40Cd)	32	20	NA	0.05/0.2	
Argobond	616/735	23	—	30	12	12.3	0.075/0.2	
Cadmium-free 'Silverflo' Range								
67E	705/723	67	(DIN L.Ag67)	NA	NA	NA	0.05/0.15	Formerly 'G6'. White in colour, suitable for brazing silver etc.
60	695/730	60	AG13	27	NA	10	0.05/0.15	Cadmium-free substitute for Easyflo 1 and 2. Tin-bearing.
55	630/660	55	AG14	26.5	NA	11	0.05/0.15	
45	680/700	45	—	26.5	—	11.5	0.05/0.15	Short melting range. Useful when brazing heavy components.
452	640/680	45	—	26.5	NA	12	0.05/0.15	Tin-bearing. Lower working temperature than Silverflo 45.
44	675/735	44	(DIN L.Ag44)	35.5	NA	12	0.05/0.2	Good fillet-forming.
43	690/775	43	AG5	26	NA	11	0.05/0.2	

(continued)

JOHNSON MATTHEY METAL JOINING LTD. *continued*

Alloy	Melting range		Spec				Impurity	Notes
40	650/710	40	AG20	29	NA	10	0.05/0.2	Tin-bearing.
38	660/720	38	—	28	NA	11.5	0.05/0.2	Wide melting range, low solidus.
34	630/730	34	—	NA	NA	13	0.05/0.2	Formerly D3. Short melting range; useful for step brazing.
33	700/740	33	—	35	NA	14.5	0.05/0.2	
30	695/770	30	AG16	33	NA	9	0.05/0.2	Tin-bearing. Fillet-forming.
302	665/755	30	AG21	30	NA	9	0.05/0.2	Fillet-forming.
25	700/800	25	(DIN L.Ag25)	27	NA	11.5	0.075/0.2*	Formerly C4.
24	740/800	24	—	30.5	NA	10	0.075/0.2*	Silicon-bearing. Unsuitable for steel under shock-load conditions.
20	776/815	20	(DIN L.Ag20)	21.5	NA	9.5	0.075/0.2*	
18	784/816	18	—	30.5	NA	9.5	0.075/0.2*	Formerly B6.
16	790/830	16	—	33	NA	11	0.075/0.2*	High, but short melting range.
12	820/840	12	(DIN L.Ag12)	26.5	NA	10	0.075/0.2*	Silver-bearing alternatives to the common 'spelter' alloys. Can be used on carbide, steel, and copper. SF1 contains silicon.
4	870/890	4	—	24	NA	10	0.075/0.2*	
1	880/890	1	—	23	NA	10	0.075/0.2*	

*Max. gap to 0.25 with copper.

Special Purpose Alloys

Alloy	Melting range		Spec					Notes
Argobraze 49H	680/705	49	AG18	32	NA	NA	—	Manganese/nickel, for brazing tungsten carbide.
Argobraze 56	600/711	56	—	30	NA	10.3	—	Nickel/indium, for use on stainless steels.
Argobraze 50	639/668	50	—	32	NA	NA	—	Nickel-bearing. High corrosion resistance. Contains cadmium.
Easyflo No. 3	634/566	50	AG9	32	20	13.3	—	Nickel-bearing. For carbide tool tips. Short melting range. Contains cadmium.

Copper-phosphorus Alloys

Alloy	Melting range		Spec					Notes
Silfos	644/700*	15	CP1	41	10	NA	0.04/0.2	200°C max, service temperature.
Silfos 5	640/710*	5	CP4	41	7	NA	0.40/0.2	Rather rough fillets.
Silbralloy	644/740*	2	CP2	32	5	NA	0.40/0.2	For electrical work and on domestic hot water systems.
Silfos 6	644/718	6	—	30	4	NA	0.40/0.2	
Copperflo	714/810	—	CP3	31	1-2	NA	0.40/0.2	

*The liquidus is rather higher, but 95% of the alloy is liquid at this temperature.

THESE ALLOYS MUST NOT BE USED ON STEEL OR ANY ALLOYS CONTAINING NICKEL. See p. 37.

(C) – THESSCO LTD. (Formerly Sheffield Smelting Co., Ltd.)

Cat. No.	Melting Range °C	% Silver	BS 1845 No.	Tensile Strength ton/sq.in	Joint Gap mm	Remarks
Cadmium Bearing						
MX20	620/640	50	AG1	28.7	0.025/0.10	Free-flowing general purpose alloys.
MX18	630/640	48	—	28.5	0.025/0.10	Close control of joint gaps needed.
MX12	610/620	42	AG2	28.7	0.05/0.10	
MX10	595/630	40	AG10	32.1	0.05/0.15	
MX8	605/650	38	—	31.9	0.05/0.15	General purpose alloys used widely in induction.
AG3	605/650	38	AG3	31.9	0.05/0.15	
MX4	610/670	34	AG11	29.7	0.075/0.2	Lower silver content in this range reduces cost. Increased melting range forms larger fillets and enables wider joint gaps to be bridged. Care needed to avoid liquation problems.
MX0	600/690	30	AG12	29.9	0.075/0.2	
LX18	605/710	25	—	29.9	0.075/0.2	
LX16	610/720	23	—	30	0.075/0.25	
LX13	605/765	20	—	24	0.075/0.25	
LX8	700/780	15	—	24.9	0.075/0.25	
Cadmium-free						
H1	690/735	61	AG4	24.3	0.05/0.15	
M14	675/735	44	AG15	32.4	0.075/0.15	
M13	690/770	43	AG5	30.6	0.075/0.2	
M0	680/770	30	AG16	32.7	0.075/0.2	
L18	700/800	25	AG17	22	0.075/0.2	
L17	740/810	24	—	31	0.075/0.2	
L13S	690/810	20	Silicon bearing	32.5	—	For applications where substantial fillets are needed.
L7	810/850	14	—	34	0.05/0.15	High temperature short melting range.
L3	830/860	10	—	33.8	0.05/0.15	High temperature short melting range.
Tin-bearing Cadmium-free						
M25T	630/660	55	AG14	27.3	0.05/0.12	Alternative to MX18 and MX20.
M10T	650/710	40	AG20‡	26.4	0.025/0.15	Alternative to MX4, AG3, or MX8.
M0T	655/755	30	AG21‡	25.3	0.075/0.2	Alternative to MX0 and M0.

‡Subject of Brit. Patents 1436943 and 1532879. Endorsed 'Rights of Licence'.

(continued)

THESSCO LTD. continued

Other Alloys

H12	MP780	72	AG7	30.6	—	Silver/copper eutectic.
MX20N	635/655	50	AG9	31.8	†	Nickel-bearing for brazing carbides.
M19MN	680/705	49	AG18	28.3	†	Nickel-manganese bearing. Cadmium-free, for carbides.
H7TN	730/760	67.5	—	21.2	0.075/0.2	Corrosion resistant alloy for stainless steel
H0T	600/720	60	AG6	30.6	†	Copper-silver-tin. Zinc and cadmium-free.
Silbraze	885/890	1	—	31.7	0.05/0.125	General engineering and structural work.

†Normally used as foil.

Phosphorus-bearing Alloys (not to be used on ferrous or nickel alloys. **See p 37.**)

Phos 15	645/700*	14.5	CP1	42.5	0.05/0.125	200°C maximum service temperature.
Phos 6	645/700*	6	—	30.8	0.05/0.125	
Phos 5	645/710*	5	CP4	30.0	0.05/0.125	200°C maximum service temperature.
Phos 2	645/740*	2	CP2	28	0.05/0.125	Mainly for electrical work and domestic hot water systems.
Phos 0	710/730*	0	CP3	32	0.05/0.125	

*The true liquidus is somewhat higher, but 95% of the alloy is liquid at this temperature. See General Notes on page 17.

Appendix II

Data on Fuel Gases

Heating Value (Calorific value) Heat is a form of energy and can be stated in any 'energy' units. The classical form is the British Thermal Unit (BTU), which is the amount of heat required to raise 1lb of water through 1°F. The Centigrade Heat Unit (CHU) is the same but related to the degree centigrade; 1 CHU=1.8 BTU. The Gas Board measure supplies in THERMS=100,000 BTU. The alternative measure of heat in SI.I Units is the JOULE, but this is rather small for engineering usage (1 BTU=1055 Joule) so that KiloJoule (kJ) or Megajoule (mJ) must be used.

The RATE of heat consumption can be measured in BTU/hr (or /min, /sec, etc), kJ/hr, or in KILOWATTS; 1 KW=3410 BTU/hr=3.6 mJ/hr. To give some idea of 'scale' a medium sized brazing torch develops about 6KW, and the largest burner likely to be used in any brazing operation is about 50KW. A 2-bar electric fire develops 2KW.

The heating values of the three common gases are as under (Methane is 'North Sea Gas' and this varies slightly from area to area.) Comparative figures are also given for 'Paraffin' (Kerosene).

Fuel	BTU/ft^3	BTU/lb	kW/m^3/hr	Air required Vol/Vol	Cu.ft/lb Liquid	Characteristic Heat, BTU/ft^3
Methane	995/1000	—	9.9-10	9.52	—	105†
Butane*	3200	21,150	32.96	30	6.51	106†
Propane*	2500	21,500	25.75	23	8.6	108†
Kerosene‡	—	20,100	—	180 ft^3/lb	—	111†

*'Commercial' gases as supplied by retailers.
†Heat release per cubic foot of optimum air/fuel mixture.
‡'Commercial Paraffin', for comparison.

There is very little difference in the maximum flame temperature developed by the three gaseous fuels (about 1980°C) but kerosene is slightly higher. The 'self-ignition temperatures' of propane and butane are about the same, around 500°C; methane is rather higher (700°C), and kerosene much lower (300°C), but these figures do depend considerably on outside conditions. (They assume that both the fuel and the air are at the same temperature, and that the air/fuel ratio was within specified limits.) The three fuel gases are all fairly sensitive to air fuel ratio, the upper and lower limits of combustion being as under.

Fuel	Methane	Butane	Propane
Lowest % Gas in air/fuel mixture	5.3	1.9	2.4
Highest % Gas in air/fuel mixture	14.0	8.5	9.5
'Theoretically correct'	9.5	3.2	4.2

LIQUID PETROLEUM GAS (LPG)

Though some small 'disposable' gas containers are charged with a mixture, most supplies are either commercial butane or commercial propane. Both are obtained from the refinement of crude oil, but some is derived direct from 'gas wells'. Methane is basically a 'Natural Gas' and cannot normally be stored in liquid form.

LPG must be stored under pressure if it is to remain as a liquid. The normal pressures used are 25 lb/sq.in for butane and 100 lb/sq.in for propane. The actual pressure in the storage bottle will depend upon its temperature, the following table giving the approximate figures.

Temperature °F	32	40	60	80	100
Butane, lb/sq.in	17	20	27	40	60
Propane lb/sq.in	68	80	100	130	175

Neither gas will 'freeze solid' at any temperature likely to be reached outside the laboratory, but if the temperature of butane falls to 14°F (−10°C) the pressure within the storage bottle will fall below atmospheric, and no gas can be drawn off. This is important, as when gas IS drawn from the cylinders the latent heat of vaporisation must come from the gas, and, with butane especially, an excessive rate of withdrawal can cause a starving of the appliance or burner. For this reason the familiar gas bottles are each 'rated' for a maximum continuous rate of withdrawal. These are given in the table below, but when brazing – which normally involves intermittent usage of gas – the rates can be considerably increased. The 'nominal size' of the container is the liquid gas content when full, in kg (lb).

Nom.	Size	3.9(8½)	4.5(10)	7(15½)	13(29)	15(33)	19(42)	47(104)
Butane	lb/hr	—	0.78	1.06	—	1.52	—	—
	cu.ft/hr	—	5	7	—	10	—	—
	BTU/hr	—	16,000	22,400	—	32,000	—	—
Propane	lb/hr	1.16	—	—	2.3	—	2.9	5.2
	cu.ft/hr	10	—	—	20	—	25	45
	BTU/hr	25,000	—	—	50,000	—	62,500	112,500

It is clear that despite the higher heating value of butane gas the 'heat rate' for comparable sizes of cylinder is much greater for propane than for butane. There is yet another consideration which favours the former gas when using self-blown torches as opposed to those with a separate air supply. The full heat rate of the burner can only be obtained when working at a fairly high pressure. Regulators supplied for use with these torches (e.g. those by Primus- Sievert) can be adjusted to one of three pressures – 28, 42 and 56 lb/sq.in – depending on the type of flame required. Clearly the normal cylinder pressure of butane is sufficient for operation only on the lowest of these gas rates. It follows that with *self-blown* torches propane is the only fuel which can give the best from the torch.

Gas Regulators A different regulator is needed for each type of LPG, because the cylinder connection is different. There is thus no chance of connecting the 'wrong' type of regulator. For each gas regulators can be had for 'high' or for 'low' pressure. The former will deliver gas at from 10lb/sq.in up to 56 lb/sq.in. The latter are usually set at 11 inches water gauge for butane and 14 inches W.G. for propane; this difference means that for a given air-blown burner nozzle the heat rate (in BTU/hr) will be the same for each gas. Self-blown torches require the high-pressure unit, air-blown ones the low. In selecting the regulator you must have regard to the gas consumption of the burner; this is especially important with low pressure regulators, as some of these are designed for use with small domestic appliances using very little gas.

Self-blown torches are almost always rated at the MASS of gas used − e.g. 9½ oz/hr or 275 gram/hr − and this can be recalculated in terms of volume from the table above. Air-blown torches may be rated in heat units/hr, which can also be so converted, but it is perhaps prudent to ensure that the regulator will pass 'as much gas as you can burn' − you may buy a larger nozzle later. This maximum gas rate will be determined by your *air supply*. If your blower delivers (say) 8 cu.ft/min − 480 cu.ft/hr − this can burn (from the table above) 480/30 cu.ft/hr of butane or 480/23 cu.ft/hr of propane; 16 and 21 cu.ft of gas/hr respectively. You should allow a margin on this, as you may wish to 'burn rich' − i.e. with less than the theoretically correct air/fuel mixture. With all air-blown torches the compressor or blower size fixes the maximum flame you can use, whilst with the self-blown type it is the size of the nozzle and the gas regulator which limit this.

If more than one burner is to be used from a single cylinder, or if you wish to use two small cylinders in parallel to supply a large torch, you should consult the local LPG dealer. There are special connectors for both purposes, and it is not safe to rely on home-built connections, even for temporary purposes.

Take care, especially if you have both low- and high-pressure services in use, that the correct hoses are connected. Most are marked with the safe working pressure. Take particular care with HP hoses, as even a small puncture will release a great deal of gas. Neither propane nor butane is 'poisonous', but both are considerably heavier than air and pools of gas can collect in low places, just waiting for a spark to cause an explosion. Hose failure devices will cope with a *ruptured* hose, but are usually too insensitive to react to leaks. The system can be checked very easily by using soapy water at all joints or suspect places; any leak will be revealed by local frothing.

As already discussed, given the right type of burner any of the three gaseous fuels can be used for any type of brazing. Butane is clearly limited somewhat by its storage pressure, but I have used both butane and propane with both self-blown and air-blown torches quite satisfactorily. The only limitation on the former arises from the size of my butane cylinder with its limited gas discharge capacity. As to North Sea Gas (methane) this is almost universally employed by industrial firms for brazing, and (provided that the service meter is large enough) will be as effective as LPG. An air-blown torch will, however, be needed as the supply pressure is only 8-11 inches water gauge.

Appendix III

Safety when using Cadmium-Silver Alloys

The following article is reproduced, by courtesy of the Editor, from the *Model Engineer*, 20th November, 1981. It was written following reports of the sudden death from cadmium poisoning of a worker who was brazing with cadmium-bearing alloys using an oxy-acetylene torch. It should be noted that the data given in Tables I and IV refer to the 1977 revision of BS 1845. Up-to-date figures may be had from the table on page 122.

THE ARTICLE by Mr. Boulden in issue 3662 is timely. It adds materially to the observations made in 'Postbag' by various writers in issues 3645, 3649, 3657, and 3660. Had the precautions which he recommends been observed the tragic death reported in one of these letters might have been avoided. However, his suggestion that the atmospheric pollution in the workshop be checked may be difficult for some readers. Further, if the figures which he quotes are used without discretion the result of such a calculation may well appear to be so unrealistic as to lead the reader to suppose that the magnitude of the risk is equally so, especially if much silver soldering has been done without apparent ill-effects. It may help if the matter is taken a little further and the following notes may give a clearer understanding of the problems.

CADMIUM

Cadmium is one of a group of metals characterised by low melting and boiling points – at 321°C and 767°C respectively. At this latter temperature, in air, cadmium vapour readily oxidises to form particles of cadmium oxide (CdO) which is toxic. (Cadmium vapour is itself toxic and a 'mist' of condensed vapour equally so; any cadmium vapour which is not oxidised will appear ultimately as a very finely divided powder in the atmosphere. This, too, is highly toxic). Zinc is another metal from the same group as cadmium, with melting and boiling points of 419°C and 907°C, which is found in brazing alloys. This too is toxic, but not to the same extent as cadmium.

To complete the picture, copper has a melting point of 1083°C and boils at about 2300°C, whilst silver melts at 961°C and boils at about 2000°C. All these boiling points lie above the melting point of the brazing alloys containing cadmium, but it must be borne in mind that all liquids can form a vapour from surface evaporation and that the melting point of the alloy is fairly close to the boiling point of both cadmium and zinc, so that measurable quantities of both metals can evaporate under brazing conditions. Silver is added to the conventional copper/zinc brazing alloy for a number of reasons, the chief of which are to reduce the melting point of the alloy, to improve the capillary action at the joint, and to make the alloy 'wet' the parent metal easily. The addition of cadmium serves similar ends. It improves the fluidity still further and has the added merit of being cheaper than silver. In addition, the flow point can be reduced, thus effecting an economy in fuel gas. The advantages of cadmium-bearing alloys are considerable, and it is most unfortunate that these must be offset by the hazard of increased toxicity – especially when used by amateurs. (And we must accept that most model engineers *are* amateurs, no matter how many boilers they may have made.) So, let us examine this in more detail.

THE NATURE OF THE HAZARD

We need not consider the effect (chiefly on the kidneys) of a continuous exposure to relatively low levels of cadmium oxide; this may concern the industrial user, but model engineers use their brazing equipment only on occasion. Even then, by industrial standards the amount of alloy used is small. We are concerned only with the immediate effect – in medical terms the 'acute' rather than the 'chronic' effects. At the upper extremes of concentration the oxide can cause death after as little as 20 minutes exposure. At a lower level – say 5 milligrams per cubic metre – exposure over a period of 8 hours can be lethal. As would be expected, the lower the concentration the less the risk, and a Threshold Limit Value ('TLV') of 0.05 mg/cu.metre has been set to which it is believed that nearly all workers may be exposed repeatedly over a 40-hour working week without adverse effect.

It is important to appreciate that the fixing of the TLV is a matter of judgement by the authorities concerned. A 'safety factor' would be incorporated to allow, for instance, for the fact that the material has only been in use for about 30 years, so that not all the long-term side effects may be known. A most important point which must be noted is the fact that the TLV for cadmium oxide is a 'ceiling' value, which must *not be exceeded at any time*; it is NOT an average over the working day. (Such a limit is called a 'Time-Weighted Average' or 'TWA'). For some poisonous gases – carbon monoxide is an example – the body can cope with short exposures which might be lethal if prolonged, provided the *average* level of exposure is below the TWA. This is NOT the case with cadmium; even a brief exposure above the threshold will have ill effects.

Breathing in the toxic fumes in sufficient concentration results in an acute inflammation of the respiratory system, which may be fatal after a single exposure. The symptoms may not appear immediately; a 'latent' intervals of some hours – up to 24 – is followed by the patient 'feeling poorly' with a fever, chest tightness and pain, and an irritating cough and shortness of breath, which may become severe. Immediate medical attention is imperative, and any *believed* to have been exposed to an excessive concentration, through accident or sheer carelessness, should be set under medical observation for at least the following 48 hours, to be sure.

TABLE I

BS 1845		Mean Content %			Melting Range °C		Vernacular Names
	Ag	Cu	Zn	Cd	Solid	Liquid	
Ag. 1	50	15	16	19	620	640	Easyflo No. 1 MX20
Ag. 2	42	17	16	25	610	620	Easyflo No. 2 MX12
Ag. 3	38	20	22	20	605	650	Ag. 3
Ag. 9	50	15½	15½	16	635	655	Easyflo No. 3 MX20N
Ag. 10	40	19	21	20	595	630	DIN Argoflo MX10q
Ag. 11	34	25	20	21	610	690	Mattibraze 34 MX11
Ag. 12	30	28	21	21	600	690	Argoswift MX0

CADMIUM CONTENT OF ALLOYS

Having described the effects — how to avoid the risk? Clearly, by keeping the atmospheric pullution well below the threshold limit value. To do this we must 'know our enemy' and Table I gives the vital statistics of the alloys containing cadmium listed in BS 1845. I have also given some of the 'vernacular' or trade names for ease of identification. There are a few alloys on the market manufactured to continental 'DIN' standards, but the cadmium content of these will be of the same order as those in the table.

The 'Mean Content' is the centre of the permitted range under the British Standard and commercial alloys may vary a little, but not more than 1% up or down. The temperatures shown as 'Solid' and 'Liquid' refer to the 'Solidus' and 'Liquidus' of the particular alloy (see page 18).

AMOUNT OF EVAPORATION

First let me clear up one point. The figure of 0.05 mg/cu. metre refers to the amount of cadmium present, whether as oxide or not, so there is no need to work out how much is oxide, which is a relief! The figure of 2% of the cadmium in the alloy quoted by Mr. Boulden is an extreme, only likely if the operator is very careless, or is using improper heating methods. (I shall refer to the use of oxy-acetylene later.) Table I shows that the temperature at which the alloy is wholly liquid lies 100°C or more below the boiling point of cadmium, and the 'working temperature' may be somewhat lower still — though most model engineers play safe and hold the temperature above the liquidus. Even so, as I have already mentioned, some vapour can be formed from any liquid, just as water 'steams' well below 100°C. How much depends upon how hot it is, and for how long. Some hitherto unpublished results of research by Messrs Johnson Matthey & Co. Ltd, have been made available to me and will illustrate this point.

Alloy Ag. 2 (Easyflo No. 2) was heated to the melting point and held there for varying periods. The temperature reached by the molten pool was measured and the amount of cadmium which evaporated was determined. The effect of heating

132

both directly, with the flame playing on the alloy, and indirectly, as in good brazing practice, was compared in two of the experiments. The results are shown in Table II.

The table shows clearly the results of prolonged heating – not only is the alloy given a longer time in which to evaporate, but the surface temperature also is increased. The virtue of heating the job rather than the joint, and transmitting the heat to the latter by conduction, is clearly seen.

TABLE II

Condition of the test	Temperature of the alloy, °C		% of cadmium in the alloy evaporated	
	Direct heat	Indirect heat	Direct heat	Indirect heat
Just melted	617	617	0.1%	0.1%
Melted and held for 10 seconds	730	685	1.4%	0.4%
Melted and held for 20 seconds	780	no test	2.8%	no test

(Published by courtesy of Johnson Matthey & Co. Ltd.)

These figures can, however, be used only as a guide. Several modifying factors apply when making an actual brazed joint. First, it is doubtful if even the most inexperienced model engineer would bring his work up to the 'cherry red' of the third test when using Ag. 2 alloy. Once the filler rod is seen to melt most of us would reduce the heat to a level sufficient to maintain that temperature; the flame is turned down a little. Second, it would be rare for a joint to be held molten for as long as ten seconds, let alone twenty. True I HAVE seen someone who, once the joint was made, again heated the whole lot up to melting point and 'added a little more rod to make sure'; but this is *not* good brazing practice, besides being expensive in silver solder! The third modifying factor is the most important. As soon as the alloy *does* melt most of it runs down, almost instantaneously, into the joint gap and only a narrow ring of molten alloy is exposed. That which is actually in the joint gap cannot evaporate to contaminate the workshop. In this respect the laboratory trial was more severe than our normal practice.

AN EXAMPLE

To give some idea of the figures involved, let me try to give an example, based on my 5-inch test boiler. This has 50 tubes. I doubt if the alloy was molten round each tube for more than 2 seconds – at the most – and the whole joint-filling operation took less than 10 minutes allowing for rod-fluxing, changing grip on both torch and rod, *and* mopping up sweat! Say 8 minutes to give the classical 'worst case' allowance. (For ventilation calculations it is the *rate* of evaporation which matters.)

First, to calculate how much alloy was used, and how much exposed? The tubes are 5/16 inch o.d. giving a circumference of 25mm. I arranged a joint gap (radial) of 0.001 inch – rather tight, but that makes no difference to the method of calculation – and the tube plate is 3mm thick. Thus the volume of alloy required to fill the gap is:

$$25 \times 3 \times 0.025 = 1.875 \text{ cu.mm } (0.001 \text{ inch} = 0.025\text{mm}).$$

Had I used a preformed ring an allowance of 25% above this would have been sufficient to allow for the fillet but as I was applying by rod I will assume an additional 50%, or say 1 cu.mm went to form the fillets and that these were exposed in the molten state for the 2 seconds. The Easyflo rods, 1½mm dia. x 600mm long, weigh 10gm very nearly, and the volume is 1056 cu.mm, from which we can estimate the *weight* of alloy exposed in the fillets, thus, for 50 tubes:

$$1 \times 50 \times 10/1056 = 0.474\text{gm} = 474 \text{ milligram.}$$

How much of this evaporated? Though I doubt it, I will assume that the work *did* get as hot as shown in the second case in the table – 730°C. If each joint were, in fact, molten for 2 seconds the amount of the cadmium content evaporated might be, by proportion:

$$2/10 \times 1.4 = 0.28\%$$

The alloy carries 25% cadmium, so that the total evaporated over the 8 minutes will be;

$$0.25 \times 474 \times 0.28/100 = 0.33 \text{ milligram.}$$

The threshold level value is 0.05 mg/cu.m so that to dilute the contaminant *to* this figure needs $0.33/0.05 = 6.6 \text{ cu.m of air.}$

This little example shows the procedure. The brazing of these tubes was a relatively trivial operation, but the same principle may be applied to any job. If you are aiming at a definite size of fillet, for example, then this can be treated as a triangular cross-section and the volume calculated accordingly. Remember that all these brazing alloys need reasonable joint gaps – I would have used rather more than 2 thou on the diameter for these tubes were I doing the job now. So, having done your sums, what air DO you need?

VENTILATION

The workshop I was using at the time I brazed up the boiler had a volume of some 60 cu.m, so that on the face of it the situation was quite safe. However, that is not the whole story. My brazing bench faced the wall, so that all the fumes had to get past my nose to reach the diluting air. The RATE of ventilation across the *workplace* is what matters. This can be estimated quite easily. 6.6 cu.m in 8 minutes = $6.6/8 \times 60 = 50$ cu.m/hr. Note that had the work been done faster, then a greater air flow would have been needed. In the event, the heat output from the burner would almost certainly be sufficient to effect this rate of air flow – the burner itself uses about 20 cu.m of air per hour and the output is about 16kW. However, it is most unwise to rely on random ventilation and, as I have indicated, this was a small job; brazing up the foundation ring (done out of doors as it happens) used a lot more alloy. The general advice is to do ALL large brazing jobs (even with cadmium-free alloys, for zinc presents a hazard, though not as serious) out of doors, or by an open window. However, this is not always practicable; in my own case the workshop roof protects me from between 80 and 100 inches of rain a year! Remember, that if you do open the window, or work just inside the garage doors (with all petrol safely out of the way, please) this is useless if the wind is blowing *into* the building – a nasty habit it has at times. This will simply blow the CdO right in your face. For smaller jobs, and for larger ones if you intend to set up specially for them, then, the safest arrangement is to have an extractor fan arranged directly above the workplace, with perhaps a baffle to ensure that the draught will flow across the hearth, away from your face. Table III gives the output of the ordinary domestic Ventaxia fans, of the type designed for setting in the wall, with automatic shutters which open when the fan is running. The roof-mounting types provide about 85% of these figures. The 3-speed switch, available as an accessory, is well worth while, and the table gives the flow for high, normal, and low settings.

TABLE III

| Size | Air Flow (cu. metres/hour) | | |
Nominal dia. (inches)	High	Normal	Low
6	335	310	270
7	540	475	405
9	965	805	665
12	1825	1705	1515

OXY-ACETYLENE

The increased hazard arising from the use of oxy-acetylene, or oxy-anything else, has been referred to in previous correspondence. The reasons are many, of which the following are the most important: (a) The flame temperature is very high, but the heat output is trivial compared with the normal gas or paraffin blowlamp. Especially with copper, the heat flows away from the joint so quickly that the excess temperature at the actual joint is inevitable if the alloy is to flow properly. As soon as the flame is removed to apply the rod (and such a flame should never, repeat, *never*, be applied to the rod itself) the joint gap chills rapidly. Application of the flame direct to the alloy will probably boil out the zinc as well as the cadmium, and perhaps some silver as well. (b) The need for blue or green goggles means that it is difficult to judge the temperature accurately, and the tendency is to over rather than underheat. (c) The presence of free oxygen in the (recommended) oxidising flame accelerates the formation of CdO both from the molten pool and from the hot, solid, alloy. (d) The stance of the operator tends to be that of 'looking over' the work and the face is much closer to it, whereas with propane (or even paraffin) the brazier must stand well back; with my larger propane burners I always need an extension tube. This means that the risk of the fumes passing directly across the face is far greater with the oxy-fuel burner. The only circumstances in which I personally would use oxygen with cadmium alloys is when repairing a small pinhole in an already-brazed joint or when using my tiny 'Microflame' torch on very small jobs indeed. Even so, I tend to use cadmium-free alloy for such work.

The use of ANY form of electric arc heating for silver-brazing is quite unaccept-able. The almost opaque glass makes temperature judgement impossible and the secondary arc-flame is so hot that almost total vaporisation is probable.

ALTERNATIVES

As I have said on previous occasions, one must keep a sense of proportion towards the risk involved in any activity. Most metalworking involves hazards of one sort or another, and the rational approach is to apply sufficient degree of care to reduce the risk (to others as well as oneself) to acceptable proportions – a 'nil' risk if possible. This requires first, an assessment of the magnitude of the risk, and then the application of ordinary engineering common sense to the problem. In this case the solution is simple – the provision of local ventilation. If you can't work outdoors, then you must provide a fan. If you can't afford one (and they are cheap compared with the cost of your boiler – or your health) then you must change over to the cadmium-free range of alloys. Table IV gives some of these, based as before on BS 1845. As before, other alloys, to DIN standards, are available.

CONCLUSION

Finally, readers would do well to remember that all manufacturers are very willing indeed to provide data sheets dealing with their products, and especially on the safety aspects. Data on health hazards can also be obtained through the local Health Authorities. Readers should not hesitate to make use of these services, for which no charge is normally made. (Though I suggest that firms be accorded the courtesy of a stamped envelope of fair size). I have had nothing but help from such firms for over 40 years. In which connection I must express my thanks for that accorded in the preparation of this article from Mr. Gillinder, Product Manager of Johnson Matthey & Co. Ltd, from Mr. Heathcote, of their Materials Safety Bureau, and from their Technical Staff at Southgate.

TABLE IV

BS 1845	Ag	Cu	Zn	Mn	Ni	Solid	Liquid	Vernacular Names	
	Mean Content, %					Melting Range, °C			
Ag. 5	43	37	19.5	—	—	700	775	Silverflo 43	M13
Ag. 13	60	26	14	—	—	695	730	Silverflo 60	
Ag. 14	55	21	22	2% Tin	—	630	660	Silverflo 55	M25T
Ag. 15	44	30	26	—	—	675	735	Silverflo 44	M14
Ag. 16	30	38	32	—	—	680	770	Silverflo 30	M0
Ag. 17	25	41	34	—	—	700	800	Silverflo 25	L18
Ag. 18	49	16	23	7	4.5	625	705	Argobraze 49H	